WHAT IS TIME?

What
Is
Time?
An
Enquiry

Truls Wyller

REAKTION BOOKS

Published by Reaktion Books Ltd
Unit 32, Waterside
44–48 Wharf Road
London N1 7UX, UK
www.reaktionbooks.co.uk

First published in English 2020
English translation copyright © Reaktion Books 2020
English translation by Kerri Pierce

This book was first published by Universitetsforlaget, Oslo
under the title *Hva er Tid* by Truls Wyller
Copyright © Universitetsforlaget 2011

This translation has been published with the financial support of NORLA
Additional support from the Norwegian University
of Science and Technology

Printed and bound in Great Britain
by TJ International, Padstow, Cornwall

A catalogue record for this book is available from the British Library

ISBN 978 1 78914 236 5

CONTENTS

INTRODUCTION: WHICH TIME? 7

ONE

THE CLOCK AND ITS PAST: ON TRADITIONAL
AND MODERN TIME CONCEPTIONS 18

TWO

PHYSICS TIME: EINSTEIN 43

THREE

PHILOSOPHY'S TIME:
BERGSON, HUSSERL, HEIDEGGER 62

FOUR

RECURRENCE TIME IN LIFE, RELIGION,
HISTORY AND LITERATURE 86

FIVE

IS THERE ANY TIME OUT THERE? 106

SIX

THE HUMAN *NOW* 126

REFERENCES 142
FURTHER READING 146
INDEX 149

INTRODUCTION:
WHICH TIME?

Have you ever attended a school reunion? Then perhaps you, too, have had a strange, intense experience. One by one old friends are introduced to you and, if you did not know at first whether you would recognize them, you happily realize that you do, at least after a couple of seconds. You realize they are still the same people; soon the conversation flows as it did in the classroom or playground. But you see faces that have changed and hear about lives after school that obviously influenced the attendees' personalities in new and sometimes startling ways. Suddenly, you perceive an entire life's course, clearly outlined back into childhood, and rather more indistinctly contoured towards a possible old age. Then perhaps you, like I, feel overwhelmed by time's passage.

Who has not in their life's wandering been amazed by time's enigmatic reality? This amazement can strike at any time, for time pursues us constantly, and we can never know when it will decide to tap us on the shoulder to remind us of its existence. But there are also certain situations that particularly invite reflection over time's essence.

In considering a photograph, we experience the way time seems to slice into life. You sit paging through a photo album together with one of your children or grandchildren: 'There you were only six weeks old ... and there's your first day of school.' At the same time, the child or grandchild at your side has managed to become an adult – just as you were when the photo was taken. You lose yourself in wonder at something that no longer exists, but that still has an unmistakable reality in the photograph before you. It is not strange that photography has unleashed philosophical reflections about time and transience, by the French literature and image theoretician Roland Barthes (1915–1980), among others. To contemplate a photograph is to stare directly into something bygone, almost like seeing a spectre, he writes.[1]

Yet what exactly fascinates us about 'time's passage'? Life is full of things that come and go; we are not dumbstruck every time we see someone walking down the road. We know, first off, that time is all-encompassing, that it is an inescapable companion. This reminds us that all things are finite and that death is an existential fact shared by all who live. In the second place, we are often captivated by the unobtainable and anyone who has sought to grasp time has felt it twist away with a teasing smile. Perhaps 'time's passage' is simply a way to attract attention, only to reveal in the next instance that there is, in fact, no 'passage' at all.

For many years, a well-known Norwegian insurance company, Gjensidige, has used this very idea to capture attention: 'Tiden går. Gjensidige består.' Time passes. Gjensidige remains. A better slogan would be hard to come by; through text and images it has become a part of the Norwegian collective unconscious. Every time we hear it, we see time's passage symbolized for us by the massive, dark shape of a nightwatchman steering his course through the dark. In this way, the slogan instils security.

As surely as time passes, Gjensidige Insurance will remain. It seems impossible to cast these two truths in doubt.

Yet if we pause to consider, if we take the complainant's corner, we do that very thing. We cast them into doubt. Gjensidige will not always remain; at any rate, time hardly passes. Time is no typical, portable object and a colleague of yours will never remark: 'On the way to work this morning, I saw a fox by the roadside, then I saw an eagle in the sky, and, hey, just down the street I saw Time turning the corner.' Unlike a fox or an eagle, time neither comes nor goes, neither on earth nor out in the cosmos. All movement happens in time, and it gives us a certain idea of movement to say something takes such and such time, but time itself cannot very well exist within a time.

Nonetheless, we never perceive things in movement without experiencing time as an attendant 'something'. So, if time does not literally come and go, it can still be a framework around all that does. It is not, however, a concrete, tangible framework and we approach time's reality more closely when we think of it as a horizon that surrounds us wherever we may go, without us being able to touch it. If time is as omnipresent as water is for a fish, it is also as distant as a rainbow to a curious child. Perhaps it was this blend of familiarity and fleeting distance that led the philosopher and church father Augustine (AD 354–430) to exclaim in his autobiography *Confessions* that, so long as no one asks him what time is, he knows, but as soon as someone asks him that question, he no longer does.[2]

Upon closer inspection, time threatens to disappear, and it is easy to sympathize with the Portuguese author Fernando Pessoa (1888–1935):

What is this thing, then, that measures us without measure and kills us even though it does not itself exist? It is at moments like these, when I'm not even sure that time exists, that I experience time like a person, and then I simply feel like going to sleep.[3]

IF A MODERN author can experience time 'as a person', it is not at all strange that people did exactly that in older, so-called archaic societies. Long before the general concept of 'time' played a practical role in human life, people related to various time deities.

The Greek Chronos represented 'origin time', while other gods have stood for time's presence in certain actions and natural occurrences. In Roman mythology, we find Janus, the 'threshold' or 'transition god' with two faces: one turned to the past, one to the future. We also have the agriculture and harvest god Saturn, who devours his own children, a divine concept with something to say about time's destructive power: that it submits everything to the law of change, which means all that remains will one day perish. Just as disagreeable is the Indian god Vishnu, who in the Bhagavad Gita epic swallows the living whole and chews men to pieces so that only bits in his teeth remain.

For societies with such gods, time's existential dimension was inextricably bound up with mythology, and reflecting upon 'time' was the same as reflecting upon specific aspects of a shared religious conceptual world. This is no longer true, but that does not mean that a modern consideration of the time phenomenon must limit itself to only one of these aspects.

I myself have long been fascinated by time's physical, philosophical, existential, social and historic reality, and in this book I have attempted the most possible comprehensive discussion of the question 'What is time?' This has meant tackling themes belonging to different disciplines, with the distinction between the natural and human sciences drawn as the principal division. Today, we distinguish not only between man and god, but between natural laws and moral and juridical ones. That was not the case in archaic societies, where the gods who determined human lives also explained the forces at work in the physical cosmos. For time's part, the calendar's and the clock's histories can serve to illustrate the evolution away from such all-encompassing cosmologies.

Human and Scientific Time

For as long as man has had a time consciousness, he has seen time's passage in relation to the change of days and seasons. The visible rhythm of heavenly bodies has defined the rhythm of time, and many archaic calendars were conceived as imperfect representations of the Sun as time's perfect, cosmic source. This concept had natural scientific, religious and political dimensions. Based on Babylonian astronomy, the ancient Egyptian calendar made it easier to prepare for the Nile's periodic flooding. Simultaneously, the calendar marked a series of religious festivals and holy days, usually tied to seasonal activities like harvesting, hunting and fishing. The fact that these demarcations regulated the lives of large numbers of people furthermore presupposed the existence of an extensive, politico-religious power centre.

Ancient Athens did not form this kind of power centre, and the Athenians of Socrates' and Plato's day did not have a cosmic calendar. Rome, however, would become the world centre. People to this day rely on the natural periodization of the Julian calendar from Julius Caesar's time,[4] meaning that we are still witnessing the consequences of Roman world domination. Meanwhile, what *has* changed is the relationship between science, religion and politics. Following the emergence of a new natural science in the Renaissance, man could no longer regard the heavenly bodies' perfect forms as the source of norms and values. And after modern notions of popular sovereignty and Enlightenment ideas arose, the Church was no longer recognized as a political entity with the power to direct people's activities according to a calendar of religious holidays.

For the part of the natural sciences, the clock would particularly further the periodization of the world's course. Modern clocks regulate people's daily activities in the same way that, in previous eras, calendars regulated yearly, monthly and weekly activities. These clocks, like the calendar, originate in ancient mathematics and astronomy, and their division into hours, minutes and seconds stems from the Babylonian sixty-number system (and their division of the circle into 360 degrees). But despite these mathematical divisions, as we shall see in the next chapter, it is a long stretch from societies with isolated sundials to modern society's thorough regulation by the watch's minute and second hands.

This regulation ensures we have become accustomed to an increasingly detailed and stabile quantification system that fails, meanwhile, to capture all the qualitative aspects of our time consciousness. As such, the modern natural sciences have resulted in two different ways of apprehending time. We have clock time for all that occurs in the physical world's inanimate material and we have experiential time to denote people's life engagement, for which dead material has no counterpart. The constellation of these two time concepts is a steadily reoccurring theme on the following pages, especially centred on one problem: the potential presence of the past and the future in the now.

If sunlight blinds us when we step outside in the morning, we have fallen victim to the Sun's eight-minute-old radiance, which will never again be physically present to us. Yet with people's relationship to their own individual or collective past, the case is different. Suddenly, we blush at the thought of something that happened to us long ago, and after significant social changes, the 'battle for history' can become a hot topic of public debate. Our relationship to the past has immediate consequences for our relationship to the present and to the future. We can say the same about political constitutions – in the U.S., for example, of the Founding Fathers' ability to obligate current and future generations.

A central issue in the discussion of these three time dimensions surrounds the phenomenon of *change*. That things change overall has always been a source of wonder: first nothing is there, then something exists for a time and then it seems to slides out of existence again. This can seem difficult to grasp, especially when it comes to human birth and death, but also in the case of less extreme changes. Picture again class reunions and similar get-togethers. It feels odd that the same person who stands here now with thin hair and a careworn face once had a full mane and ruddy cheeks. No equal astonishment greets one person having thin hair and another person having a full head of it. So what does it mean that one and the same thing – and not simply two distinct things – has different properties at different times? Is there a place for any actual change in the natural sciences?

When I address such questions, I also indirectly address worldviews and human perspectives that touch on the human soul's cosmic loneliness in a physical universe. Are we, when it comes down to it, simply cogs in an eternal, mechanical clockwork, or do we represent, within our lived past and present horizon, a time that is unique in the universe? The most important thing in this context is perhaps not to distinguish between two time concepts, one for inanimate nature and one for human self-understanding. In any case, we should look closer at their relationship in order to see whether one time concept will prove more fundamental than the other, or if both are necessary in order to discuss time at all.

The Book's Contents

Before plunging into the question of what time is, we must dwell a moment on another question: does the uniform phenomenon 'time' exist at all? If it does not, the first question is utterly meaningless. What if human life's experienced time and physics' experience-neutral time simply have the word 'time' in common, much in the same way that boxing up a present and boxing someone's ears both use the word 'box'? What if the various modern sciences and disciplines each have their own time concept without any interesting contact points? Indeed, what if every natural phenomenon, every society, every activity or perhaps even every person has their own special time?

The clock's dominance in modern society makes it rather too simple to dismiss these questions. Every time series in the universe can be measured with a synchronized watch, we like to think, which means it is permeated by a single, homogeneous time. But just how different life can appear in premodern, more or less clockless societies. I encourage the reader to imagine this in Chapter One, in light of holiday experiences far away from any city's hustle and bustle. I also look closer at some historical prerequisites for clock time's social dominance. Towards the end of the chapter, I briefly discuss to what extent uniform time exists overall. I do not present a definite answer, but I would not have proceeded with further chapters had I thought it was meaningless to reflect generally on what 'time' is – as the last century's great physicist Albert Einstein did, among others.

By the end of the 1800s, clock time had made it difficult to clarify what exactly it means that things happen simultaneously, both within the sciences and within society. In Chapter Two, I try to show how Einstein approached this problem in a way that revolutionized the physical time concept. Like the sixteenth- and seventeenth-century physicists Galileo Galilei and Isaac Newton, Einstein attempted to develop a universal physics, and at the centre of this physics he placed the speed of light as a universal natural law. That is, light propagates itself at a speed of 300,000 kilometres per second, independent of how you move relative to it. In order to make this physics intelligible, we must, according to Einstein, 'relativize' the simultaneity concept: if two things simultaneously happen for me, that does not mean that they happen simultaneously for you.

I illustrate the critical point in this theory's presentation with the two diagrams on pages 45 and 46. Readers with relevant backgrounds will consider these diagrams to be elementary, whereas for others the thought process will seem foreign and the diagrams difficult to comprehend. I believe that dwelling on them a bit as food for thought is beneficial, but it is entirely possible to read further without full and comprehensive understanding.

Difficult to grasp, according to Einstein, is the relationship of such physical time to man's experience of existing 'now'. Within such an experience, time is not merely a series of actions that follow each other. It also perpetually encompasses a relationship between present, past and future, and Chapter Three is devoted to three philosophers who were concerned by how these time dimensions interact: the Frenchman Henri Bergson and the Germans Edmund Husserl and Martin Heidegger. Many people no doubt think each philosopher highlights essential aspects of time's role in human life. To what extent mankind's time horizon also encompasses physics' time, however, remains an open question. In the remaining chapters, we will acquaint ourselves with different viewpoints on this issue.

In Chapter Four, I first take a closer look at time's hermeneutics, a subject tied to individual and social recurrence in history, religion, art and literature. On the one hand, doing the same thing repeatedly can give one the feeling of ruling the world. On the other, it is impossible to do exactly the same thing again, since the consciousness of repeating something alters the experience of it. According to Heidegger's student Hans-Georg Gadamer, neither real history nor artwork exists entirely completed in the past, independent of our own creative rethinking. As demonstrated by the Frenchman Paul Ricoeur's narrative, poetic time theory, literary repetitions can create form and unity in otherwise disparate, episodic events. In this way, Ricoeur believes that all time stands in relation to human stories.

In order to better approach these ideas, I return in Chapter Five to the natural scientific time concept. Central to this chapter is English time philosopher John McTaggart's critical discussion of the phenomenon of change. As it turns out, it is difficult to conceive of time and change without the past and present attaining a certain reality in the now, and it is difficult to conceptualize this reality as being other than a horizon for conscious actions. If this horizon encompasses everything that physically occurs in time, it must be universal, as it seems to be among human beings, though not among animals.

Does time, therefore, not exist outside of human beings? In Chapter Six, I argue, with my springboard in people's relationship to the now, that this is, indeed, the case. That is, I turn first to poets and thinkers who have written about living in the now, where our relationship to the now has practical consequences for us. After that, I claim that this relationship also has theoretical consequences: that real, actual time series have only a practically determined, variable duration. The conclusion becomes that there is not in fact one time for conscious creatures and another for inanimate, spatial objects, but a human time that encompasses both. But let us begin at the beginning, when mankind did not have a unifying time concept.

ONE

THE CLOCK AND ITS PAST: ON TRADITIONAL AND MODERN TIME CONCEPTIONS

If you have ever holidayed at a cabin, then you have probably had a taste of the clockless existence. Let me offer my own personal experience from years back. I travelled with my family to an island at the mouth of a fjord. On the very first evening, we removed all our watches and left them in a bedroom drawer. This was before the age of the mobile phone, the cabin had no electricity and we never listened to the radio. After that, it was a matter of simply establishing a daily rhythm based on our bodily and natural moods, something that, after a brief adjustment period, functioned unproblematically.

We got up when we woke. None of us were early risers, so no one disturbed the others. Depending to an extent on wind and weather, we took our morning baths, and then we put on coffee and carried the breakfast things outside. Eventually, we became quite adept at 'reading' the clouds and the sea wave patterns, and we estimated when fishing chances would be decent. Some of us ventured out in the rowing boat to inspect the best fishing places in turn. Meanwhile, other family members repaired the roof, cleared out the shrubs, fetched water or lay in a hammock with a book.

When the sky began to darken or an ice-cold wind whipped the waves, the 'home contingent' guessed the 'fishermen' would be back soon. Then we might all take a short walk together before again brewing some coffee and eating a few slices of bread. After that, we would play some rounds of cards and walk to the island's shop to buy food. The shop is quite a distance away, and everything happened at a slow tempo, so back at the cabin we again felt hunger begin to gnaw. Some of us started washing potatoes while others heated water for a small amount of laundry. After dinner, the Sun was already sinking but still we managed a long walk along the shoreline before darkness set in. Tired from the sea air, it felt nice to creep into bed at . . . when exactly? We had stopped thinking about what the clocks in the drawer might show.

The rather unusual thing about this holiday was that it extended over several months. I received a writing grant and managed to convince the rest of the family to spend that whole autumn at the cabin. We got permission to teach the kids ourselves, and no one complained about a few school-free months. As time went on, significant changes occurred in the natural rhythm by which we lived. The evening air became clearer, the morning mist cooler, the days shorter, the east wind more infrequent and the sea often more turbulent. We also slept for increasingly longer periods – before we nearly succumbed to winter dormancy and decided it was time to return to civilization.

Of course, this story is not entirely true. But it is not too far from it either, and I trust it illustrates how different time can appear in an existence without clocks. We need travel no further than our own cabin to experience what is often called action and event time, in contrast to clock time.

During our extended holiday, the days did not consist of natural occurrences and human tasks divided into hours, minutes and seconds. Instead, they consisted of natural occurrences and human tasks – full stop. We got up whenever we woke, slept whenever we were tired, ate when we were hungry and otherwise followed natural rhythms. The day was certainly divided into these various activities: everything did not simply coalesce into a diffuse action mass. Furthermore, everything happened in a specific order: first wake up, then a morning bath, then breakfast, then fishing, then woodcutting and so on. We did not, for example, first fish and then wake up. In other words, everything happened according to a time sequence. But as long as the clock's tick did not direct our lives, this time was completely different from that which otherwise permeates modern life.

Clocks are mechanisms that run and remain unaffected by everything happening around them. If we compare other events to the clock's ticking, these events also become equal. A clock hour is an hour no matter what we determine an hour to be. And a clock perpetually repeats minutes so that every past and every future consists of precisely the same length units. As a result, both quantitatively and qualitatively, clocks make everything the same; they homogenize time. The difference between this time and the clockless time we experienced on the island is striking.

For the first few days, we naturally had a kind of clock time consciousness. 'It's probably not more than an hour until the shop closes,' I thought once, for example, as I was chopping wood. That was because closing time was our only contact with the clock's time points. But eventually 'hours', 'minutes' and 'seconds' slipped from our consciousness and the duration of something was evaluated only in relation to concrete events and actions. 'Has it been long since you came back from fishing?' I was once asked, for example. 'Yes, the Sun was just visible behind the hill, and since then I've brewed coffee twice, painted the last window frames, cleaned the shed, and finished Chapter Seven,' I replied. 'Quite a while,' in other words, but more precise than that I could not be. And no one was interested anyway.

What could be interesting was the order of the various activities, activities not reduced to qualitatively equal counting units. But no one would see the point in comparing the different activities quantitatively – discovering, for example, that painting took 7⅙ times as long as coffee drinking, or that cleaning took only a quarter of the time it took to write. Even if someone had grabbed a watch and calculated these things for fun, it would have played no role in our activity pattern and in our associated time consciousness. So, whereas clock time is distinctly quantitative, our action and event time was distinctly qualitative. Even if one even clearly takes place before the other, woodcutting and fishing are experienced as very different activities.

Now, there are some things that repeat at regular intervals on the island: the cycles of days and years. But when they influenced our activities, it did not happen mechanically, as with a clock. Air and sea temperatures change from week to week, the length of days and nights shift, and vegetation, fish and animal life vary with the season. We were certainly aware that each year is equal in length, but that knowledge hardly plays a practical role. Neither does the fact that long winter nights and short summer nights can all be characterized as having, let us say, a ten-hour duration. If we agree to return from fishing 'at sunset', it plays no practical role that the clock face marks this point in time much later in the summer than during the autumn. And as to the night itself, how can we grasp its length?

How can a time determined by human actions and sunsets and other natural events establish the length of the intervals between the concrete things we experience? How long is night when it is simply dark, when we are not active and when we are largely ignorant of what is occurring around us? We rely on qualitative divisions, such as 'when I could still see the rowanberry tree against the sky', 'after I woke from that bad dream' or 'before the birds began to sing'. Yet, the duration of the periods 'before' and 'after' are open, and if we had lived long enough without clocks, we would have lost all sense of the length of such periods. In that case, the night's time determinations would no longer be continuous, lacking 'holes', but discrete, with empty transitions between different experiences.

Furthermore, the time relationship between things that occur simultaneously, at different locations in space, would be largely undetermined. As long as we collectively organize our lives according to the clock, we know that one and the same clock strike represents all events that take place concurrently with it. Yet how can we talk about simultaneity without an unambiguous and precise time division for what people are doing at various places? We can certainly do it with reference to concrete, limited activities: 'No, I'm not able to reel in the line while I row. You do it.' One person fishes while another rows. Or: 'I'll grill the fish while you lay the table.' We do it at the same time. But what about: 'When you were finishing the chapter, was it as I was reeling in the mackerel?' First of all, no one would ever ask such a question, at least as anything but an amusing pastime. And second, how in the world could we ever know?

When I stepped into the boat, it had to be bailed first, so I had major difficulties starting the motor. After that, I sailed to the cod and saithe fishing grounds, where I waited 'a while' for the first cod, after which I took in four, and then landed a mackerel. How long all that took relative to what my wife was doing at the cabin, I had no idea and it would be difficult to resolve. Without the clock's precise counting units, any answer would be nothing but guesswork based on a vague awareness that: 'Before I returned, she was also doing this and that.' How exactly this and that might correlate to my activities at sea, no one could say. We were not terribly interested either.

Only when engaged in a common pursuit was there a point in synchronizing our activities: 'Did you see that fantastic play of colours in the sky right before the Sun disappeared?' There we had a common experience to draw on. Or: 'We'll meet at the bird tower when we're both finished with what we're doing.' In the second case, it was not necessarily important that we arrived at the same time, though when we first met it was 'simultaneously'. Aside from such instances, we did not retain any universal, all-encompassing concept of simultaneity common to our separate and very different tasks.

Let me now gather several threads from these considerations. The clock makes everything that happens in time qualitatively equivalent, so it is therefore homogeneous. Instead of different qualities, the clock registers different counting units, so it is therefore quantitative. If we further assume that the clock measures not only itself, but a 'time' common to all the world's occurrences, then this time exists independent of the various, concrete actions and events in which we engage. As such, it is abstract and universal: it encompasses natural events and actions, the intervals between them and all conceivable, continuous subdivisions of them. It also encompasses simultaneous relationships between practical, distinct activities at different points in space. With each of these points, it becomes the opposite of the clockless action and event time on the island. In other words, whereas clock time is quantitative, homogeneous, continual, abstract and universal, action and event time is qualitative, heterogeneous, discrete, concrete and local. Therefore, my time experience on the island should have significant commonalities with the time experience typical to older societies, if we are to trust historians and anthropologists.

Premodern Action Time

A society thoroughly regulated by the clock is a distinctly modern phenomenon, unknown to people in so-called premodern societies. If we modern people know anything about time experiences in these earlier societies, it is, among other things, because they exist side-by-side with our modern lifestyle – in spatially distant locations. As long as Westerners have conducted expeditions to Asia, Africa and Latin America, they have been fascinated by people's relationship to time in these foreign societies. And when someone makes a trip, they always have something to report when they return home. In our day, we are used to hearing both anthropologists and backpackers describe shocking encounters with societies that lack appreciable time control.

A 'Western man', let us call him John, is on a journey through the expansive desert and steppe regions in country X. At his hotel, he manages to obtain bus routes and has calculated that, with bus and train time, the trip will place him in city Y in four days. From there he can get a flight, so he will make it home for his son's birthday. He wakes up in time to reach the first bus at 8 a.m. At the bus stop, there are two other people, but no bus in sight. Minutes drag by, and eventually hours. One or two other people show up and by late afternoon there are seven or eight people, and still no bus. 'We still have to wait a bit,' they say when he asks. Late that evening, some settle down to sleep; others pass most of the night with board games or just with quiet conversations. John soon realizes that he will miss his appointments in the U.S., and he is so frustrated it robs him of all rest. Around 2 p.m. the next day, there are forty people – and the bus comes.

The passengers find their places, the bus fills up, and it leaves. The amusing part is that while John reacts to this, none of the country's inhabitants seems to think anything is wrong. What we only taste on holiday is their daily life. They do not live according to the clock but according to action patterns, and time is determined when one action naturally replaces another, not when the timer reaches a certain point. It is not that everyone turns up before the bus goes, but the bus goes when everyone has turned up.[1] It is this attitude towards time that turns the hair of eager aid workers grey, when people in premodern societies must be taught to heed work, school or meeting times, and to comply with deadlines for transport, deliveries, planning or building operations.

John's experience with the others' equanimity as they waited also seems typical. He became frustrated with their lack of frustration. But that lack is not so difficult to understand. Without deadlines and meeting times hanging over you, there is not always a clear distinction between activity and cessation. Without an external schedule, actions have their own, inherent tempo, which can vary from quick-paced if you are fleeing a raging rhino to slow if you are . . . waiting for the bus. There is no sharp distinction here between activity and lack of it. And not least there is the *value* one places on what one is doing. Slowness is no less valuable than haste. Indeed, why should fleeing a raging rhino be any more worthwhile than a late-night conversation?

Often, the exact opposite is true. In many societies, there is scarcely anything ruder than brushing off someone you happen to encounter after a couple of minutes 'because the rest of the day is full of appointments'. It is not socially acceptable to keep appointments at the expense of encounters with others who unexpectedly turn up. Viewed in this way, it is the quality of the individual meeting that determines the quantitative time sequence, and alone this sequence has no inherent worth.[2]

Nor do the movement of earthly and heavenly bodies exercise a single, direct influence on societies dominated by action and event time. All people at all times have naturally recognized seasonal or at least daily changes. But these changes signify something because they affect climate, animal life and crops, that is, natural events in the environment; events that in turn figure into our action patterns. They have meaning as an aspect of human life's natural surroundings, but that is it. Therefore, the day's division usually varies with changes in human activity level, with many 'hours' when a lot is going on and just a few when little is happening.[3]

The Norwegian anthropologist Anders Johansen gives us some examples of how the day is divided by actions and natural events:

The sun's positions serve, like other types of reference points, to precisely characterize the shifting states, atmospheres, and conditions for action throughout the day. Such is also apparently the case with the following time

expressions. From Malaya: 'Before the flies stir' signifies
the wee morning hours; then follows 'the heat sets in', 'the
mist dries up', 'the plough rests', 'the shadows are round'
etc. – until 'the buffaloes go to drink' and 'the children are
asleep'. The Komos in Central Africa distinguish, in similar
fashion, between phases like 'the buffaloes go to drink',
'first cockcrow', 'everyone has gone to the forest', 'the sun
is straight overhead', 'the hawk flies', 'the sun shines in
the eyes', 'sight is lost', 'time to go indoors'. A like division of the day is also reported from Madagascar: 'The
frogs are croaking', 'the eastern sky is red', 'the cow's colors
are visible'/'the diligent are waking up', 'sun up', 'the mist
falls'/'the cows go out', 'the leaves are dry', 'the cold is
vanishing' – etc. On Sumba, the day is ushered in by 'the
morning star rising', and thereafter by 'a lone helmeted
friarbird shrieks'; the morning strides on with 'the horizon is white', 'palm lines are visible', etc.; after darkness
falls, the passage of time takes the form of 'the clink of the
rice barrel', by the fact that 'the commoners' and then 'the
nobles eat', and that 'dogs and pigs are sleeping'.[4]

In such premodern societies, the movements of earthly and
heavenly bodies help to date things to the extent that they affect
people's natural surroundings, and these surroundings must in
turn be integrated into our social lives. All time, in other words,
is action time, of the type familiar from modern holiday life. The
difference is obviously that for premodern man we are talking
about more than just a sampling of another lifestyle. This *is* their
life, from birth to death, and it is integrated into a universe of
traditions, myths and cosmological ideas. Such things have deep

roots, even though almost all indigenous peoples today have contact with modern society through transport and communication means. The fundamental time experience is simply different to that to which we are habituated.

Take the distinction between what is often called an event's 'subjective', experienced duration and its 'objective', physical duration.[5] This idea is clearest in relation to a clock mercilessly ticking towards the future, independent of what we are doing. The clock's movement is also independent of any time experience, whether we think time is 'dragging' or 'racing'. This difference, however, is not always clear in a society dominated by action time. When the clock does not determine time by juxtaposing physical time periods with each other, then duration's subjective experience becomes that much more important. Sometimes it is quite simply stated that something lasts longer if it feels longer: you do not need to hurry to beat the sundown because 'when you move slowly, time passes slowly.'[6]

In such cases, speed does not decompose into a pure time sequence and pure spatial stretch (as in 'kilometres per hour'). Distance is seemingly absorbed by the movement. Although premodern people acknowledge differences between time and space, the two often run together in their conceptual world when it comes to distant regions and time lengths.

We modern clock folk are accustomed to projecting homogeneous counting units onto an indefinite past and future. Even if there are limits to how much of this universal time we take into account in our daily tasks, immediately we understand what is meant by 'in three million years' and 'two seconds after the Big Bang'. We are also familiar with cosmological and astronomical theories concerning things occurring 'now, right now' in some distant place, far out in space. We clearly distinguish as well between time and space, we relate to time and space as universal, homogeneous entities, and we have no problem imagining that something far away in space is happening simultaneously with something close at hand. How far from obvious these ideas are, however, is something we experience on clockless holidays. As we have seen, the thought that one thing is happening simultaneously with another in some other place plays no practical role. Nor does it in typical premodern societies.

It is not uncommon, in fact, that faraway things simply fall under the umbrella term 'distant', whether that idea applies to time or space. And is this really so strange? Living beings who might exist in faraway places take time to discover, and in concrete action time distance is conceived in terms of 'significant movements' away. As such, mystical figures from a far, far away past are promptly localized in a far, far away region of space. Often, distance in space will also be perceived as unified with distance in time, so that far-off ancestors somehow inhabit rocks, cliff overhangs, olive groves or lakes in remote places. None of this means, however, that premodern people did not experience any difference between time and space. Instead, it indicates that without our modern, universal concepts of time and place, the distinction between them does not play a role beyond daily life's concrete interactions.

What about the quality of life in societies permeated by concrete, qualitative action time as compared to societies permeated by clock time? Whether concerning the collective mourning of a deceased tribe member, the building of a temple, the roasting of a chicken or the hunt for a buffalo, the activity lasts 'until time is ripe' for another, not until a future time point agreed upon in the past. Under such conditions, people live more in the now and are probably more present in what they do than a person perpetually looking towards an abstract, homogenous past and a similarly abstract, homogeneous future. It is this direct presence in daily activities that modern people sometimes attempt to recreate by holidaying away from the city's hustle and bustle.

Which form of socially instilled time consciousness makes people happier, however, I have not indicated. The social sciences can hardly answer that question, and perhaps the idea of comparing life quality in different historical eras is futile. What we can determine is that people's relationship to time has been dramatically altered during the last few centuries of Western history. A society permeated by clock time is a society where clocks not only measure certain actions' duration. The clock's homogenization also invites us to consider 'time itself' as an abstraction, separate from our concrete actions *in* time. By quantifying such time, one can talk about doing so and so much per time unit. In addition, one can focus on utilizing and economizing time, an idea that is inconceivable in premodern societies.

Modern Clock Time

Relatively early in Jonathan Swift's narrative of *Gulliver's Travels*, the Lilliputians recount all the strange things they discovered in 'Man Mountain's' pockets. The strangest thing of all turns out to be a handsome chain with a round disc displaying black symbols and figures. They considered it to be some type of machine:

> He put this engine to our ears, which made an incessant noise like that of a water-mill: and we conjecture it is

either some unknown animal, or the god that he worships: but we are more inclined to the latter opinion, because he assured us, (if we understood him right, for he expressed himself very imperfectly) that he seldom did any thing without consulting it. He called it his oracle, and said it pointed out the time for every action of his life.[7]

The Lilliputians' first meeting with a clock is interesting in two ways. First, it reminds us that mechanical, machined clockworks are, historically speaking, relatively new phenomena. Second, it implies that, as a modern man, Gulliver lives his life by the clock god's sign. The second does not follow from the first, because there are many societies with clocks where the clock does not govern people's activities.

Cultures dominated by event and action time have had sundials and water clocks and other time-tellers that did very little to regulate people's activities. This fact was due in part to these clocks' unreliability or irregular functioning. A sundial, for example, is not much use at night or under heavy cloud cover. Only after the Renaissance did a mechanical clock appear that defied seasons, weather and wind and so functioned 'as a clock'. This development, which began with the pendulum clock, has led to today's digital watch. As a result, we are all familiar with instruments that measure homogeneous time quantitatively. In many places on the globe, however, people have clocks but don't use them to direct their daily tasks. The wristwatch is worn more as a status symbol, as a glittering, expensive piece of jewellery. The same can also be true in modern, metropolitan subcultures. Even though people have clocks, the lives they lead can be far removed from the clock god's sign.

This clock-life has been ardently described by the Argentine-French author Julio Cortázar:

> Think of this: When they present you with a watch they are gifting you with a tiny flowering hell, a wreath of roses, a dungeon of air. They aren't simply wishing the watch on you, and many more, and we hope it will last you, it's a good brand, Swiss, seventeen rubies; they aren't just giving you this minute stonecutter which will bind you by the wrist and walk along with you. They are giving you – they don't know it, it's terrible that they don't know it – they are gifting you with a new, fragile, and precarious piece of yourself, something that's yours but not a part of your body, that you have to strap to your body like your belt, like a tiny, furious bit of something hanging onto your wrist. They gift you . . . with the obsession of looking into jewelry-shop windows to check the exact time, check the radio announcer, check the telephone service. They aren't giving you a watch, you are the gift, they're giving you yourself for the watch's birthday.[8]

How can it be that so many modern people will nod in recognition at this description? Which psychological and social realities underlie the clock's power over our lives?

If the clock plays such a great role as regulator of modern social activities, it is no doubt connected with this society's obligated division of labour and its technical–industrial level. The industrial society, after all, will never function without work processes organized down to the smallest detail and then coordinated with other producers, such as suppliers and receivers. Scheduling, for its part, is a timetable that requires every employee in the organization to be focused on coordinating their work with activities in workplaces near and far in the areas of production, transport and trade. This industrially determined coordination is unthinkable without a drastic synchronization of human life that seems incompatible with premodern society's action time. Technically, it is certainly possible to allow customers and suppliers to wait until one feels compelled to finalize a product, or 'let the bus leave when everyone has shown up'. But a well-functioning industrial society presupposes a work-life disciplined by the clock's synchronization.

Nonetheless, the need to coordinate work processes does not fully explain the abstract, homogeneous clock time's power over modern human life. For synchronization does not itself imply that time be regarded as an independent entity, something it is possible to lack and with which one must 'economize'. In any case, it is possible to imagine a thoroughly synchronized society where people collectively control time instead of being controlled by it.

Take again the description of the island. Perhaps we found that there was much to do before our departure; some things we could do alone and some things required cooperation. While one person carried things from the boathouse, chopped the rest of the wood and gave the outboard motor a final overhaul, another could paint the window frames, throw out the rubbish and collect pretty rocks from the beach to take home. But actions like dragging the boat onto land or popping into the neighbour's garden for a farewell coffee required that we met at predetermined times. Therefore, the clocks had to be retrieved from the bedroom bureau. Yet, synchronization does not imply any time pressure: 'Let's not stress these last few days, now that the Sun is shining and everything. If the Little Man goes with you into the bathhouse, let him hang out there a bit . . . When I think about it, I estimate we need two days to get it all done.'

So long as we were all in agreement, a slow tempo did not imply poor coordination. The organizing of activities certainly put a pressure on anybody who lagged behind, but even so, the combined, planned activities would not be put under any time pressure. We synchronized our different tasks, but we did not need to join forces in order to 'utilize time'. Synchronization and 'time economics' are not the same thing. Still, it is the economization of time that characterizes foremost the clock god's influence, as summarized by the saying 'time is money'. The day's hours can even be compared to diamonds:

> Time can tear down a building or destroy a woman's face
> Hours are like diamonds, don't let them waste.[9]

Again, I am pondering how peculiar the modern time concept is. In a society dominated by action time, the very phenomenon of time economization seems incomprehensible. Some activities, like fleeing a raging rhino, require a highly charged tempo. Other activities, like singing to a sleeping child, do not. In modern societies, there are also certainly arenas in which we experience action time's logic, where time is adjusted to our activity pattern. Out in nature or at a party, we can forget time, and it seems odd to ask, 'So, did you have a beneficial experience of the sunset?' or 'Was that an effective party?' Otherwise, our life is dominated by the idea of time economics; we must get the most out of time as it is a scarce commodity.

When it comes to points in space, however, we almost never reason like that. To a Norwegian, it is fun to imagine that half the world's population could apparently fit on Lake Mjøsa, if we just stood close enough. Competitions are held for how many people can stuff themselves into a telephone booth. On a more serious level, investors want often to subordinate all the natural resources in an area to human activity, and occasionally there is talk of strengthening a nation by increasing that country's population. Yet it is extremely rare for anyone's goal to be increasing population density in the place they live. No Mjøsing wants to transport the entire world's population to Mjøsa to better utilize space. Nonetheless, it is common to want to do something every moment of now, in order to better utilize time.

An important difference between time and space is that whereas we can freely move about in space, revisiting places we once left, we cannot move around in time. What is past is gone, it never returns, and if we allow an hour to slip by, it is lost forever. We cannot return in order to 'populate' it with more human activity. This distinction between time and space explains something of the emotion behind the verse 'Hours are like diamonds' – housed in a time that will not wait. Who would bother to sing 'Metres are like diamonds'? Again, we are faced with the question: what it is about modern life that has transformed time units to rare commodities on a parallel with diamonds? Let us now take a look at what two classical sociologists have to say about this idea, first Max Weber (1864–1920) and then Karl Marx (1818–1883).

One of Weber's most famous works is *The Protestant Ethic and the Spirit of Capitalism*. Weber is concerned with the complete rationalization of life that takes place in modern capitalism, and he focuses particularly on the Protestant work ethic. In previous epochs, the wealthy strived for pleasure and opulent luxury. The capitalistic entrepreneur, by contrast, is concerned with productivity – not in order to increase consumption but rather to reinvest in consistently new productive enterprises. Capitalism's and the workers' ideals are moderation and industry, with every hour of the day maximally utilized, as Benjamin Franklin (1706–1790) suggestively formulated it: 'How much more than necessary do we spend in sleep, forgetting that . . . there will be sleeping enough in the grave.' This attitude can be traced back to the Protestant revolt against Catholicism's empty ritualism.

The Protestant work ethic characterizes, according to Weber, Calvinism and other Puritan movements in England. Within Calvinism, the individual is not chiefly part of a large religious community. Instead, the individual stands alone before God, and the path to salvation does not travel through indulgences and other external acts, but through faith and one's inner disposition. Within this construct, Weber discovers a strange duplicity. God has already predestined those who will be saved, so anxiety on that score is pointless. A person should simply live their life as befits a believer, in God-fearing sobriety. At the same time, a successful life on earth is a God-given sign that one is part of the elect. Material goods should not be pursued, but if a person heads a prosperous concern, there is joy to be had in success, both in the market and in one's relationship with God. These goals can be attained by refraining from consumption and instead demonstrating the following work approach:

> Waste of time is thus the first and in principle the deadliest of sins. The span of human life is infinitely short and precious to make sure of one's own election. Loss of time

through sociability, idle talk, luxury, even more sleep than is necessary for health, six to at most eight hours, is worthy of absolute moral condemnation. It does not yet hold, with Franklin, that time is money, but the proposition is true in a certain spiritual sense. It is infinitely valuable because every hour lost is lost to labour for the glory of God. Thus inactive contemplation is also valueless, or even directly reprehensible if it is at the expense of one's daily work.[10]

We can add that Protestantism's ideal of living life according to active time utilization has its precursors in the early Renaissance. Among the Renaissance humanists in art, religion and science, we find a dawning consciousness of the individual's cosmic isolation in a life surrounded by empty time before birth and after death. As such, it is worth filling the days we do have with meaningful activities, ideally for the sake of our eternity and to God's glory. Still, this attitude towards time exploitation is far from characteristic of religions in general. Take, for example, the admonishment children receive growing up in a society dominated by Christianity: 'Don't just sit there. Find something to do.' To this statement the meditating, Zen Buddhist master would answer: 'Don't find something to do. Just sit.'

Because we generally know the strong grip Christianity had on people's outlook in the European Middle Ages, there can be no doubt that this new attitude towards work life can be viewed against the backdrop of the Reformation and its religious renewal. Karl Marx, for his part, approached the issue in a completely different way to Weber. Marx was concerned with the capitalistic market economy's distinctive characteristics; no previous epoch had allowed for such free investment in production means, raw materials and the workforce. Earlier, these production means had been safeguarded by guilds, landowners, feudal lords, the clergy, village communities or families and clans. The capitalistic right to invest dismantled such communities, and Marx focused his attention on the market economy's free investments as a requirement for *re*investment.

In order to earn one's livelihood in the capitalistic market economy, the individual is forced to enter the competition surrounding the buying and selling of labour and other commodities. Those who do not contribute receive nothing from this market, and what one does get for a product stands in a certain relation to the amount of work that goes into it. If not, everyone would look for jobs that offered a quick pay cheque. Furthermore, one must work at least as effectively as one's fellows, something that means utilizing time at least as well as others. No one can claim, 'We produced this car in two hundred work hours. Even though the competition produced two cars in that same time, we ought to get paid for the same number of hours.' We cannot do that because what counts in the market is not the concrete amount of work time put into a concrete product, but the market's recognition of what Marx termed abstract work time.

On the one hand, such work time is indeed an abstract concept, for it takes some people longer than others to produce the same product. On the other hand, there is a social reality that ensures that wares of the same type will all appear to be incarnated, homogeneous work time. The test as to whether something counts as such work time is exchanged against money. Money becomes the anonymous market's measure for realized, homogeneous work time, which means abstract time becomes a force relentlessly imposing effective utilization on people: 'If you waste me, you risk losing your job or going bankrupt. Not only will you end up with less wealth, you will become poor and socially excluded.' Within a modern market economy, it can take tremendous individual effort to break loose from the clock god's economic grip on life.

No matter what we perceive to be the basis of modern man's relationship to the clock's abstract, homogeneous time, this time is not something we simply amuse ourselves in registering whenever we take a glance at the clock. It is a social reality that governs an appreciable amount of our lives. One can, nonetheless, also wonder whether such time even 'exists' in a world independent of the modern lifestyle. This is a question that, among others, has occupied anthropologists.

Clock Time: Discovery or Invention?

In his introductory work on social anthropology, *Small Places, Large Issues,* Thomas Hylland Eriksen discusses premodern societies that only recognize concrete action time. In such societies, time is not a scarce commodity of which we can have too much or too little or that can be measured monetarily. This fact stands in contrast to modern societies, he writes, where time is perceived as abstract clock time. As he continues:

> Time, in this kind of society, is conventionally conceptualised as a line with an arrow at the end, where a moving point called 'the present' separates past and future. This kind of abstraction is a cultural invention, neither more nor less. In a certain sense, clocks do not measure time but create it.[11]

Is Hylland Eriksen correct here? Do our watches serve simply as a means to organize modern man's activities? Undoubtedly, he is right that an abstract time that somehow exists independently of actions and concrete, natural events has not always been a social reality. Indeed, as we shall see in the next chapter, not all physicists view time this way. But does that imply that clocks are simply objects designed to keep pace and not to measure anything beyond their own rhythm? Or is that simply their task in modern society?

In order to approach this idea, we must first clarify the difference between discovery and invention. Columbus discovered America but he did not invent it. Benjamin Franklin invented the bifocal lens but did not discover it. Therefore, did the constructors of the mechanical clock invent the quantitative, homogeneous flow of a universal time, or did these clocks serve to shed light on nature's own time? This issue raises fundamental, philosophical questions to which I will give no definite answer. Let me instead present a couple of alternative viewpoints.

Hylland Eriksen emphasizes the significant difference between modern and premodern time conceptions. Anders Johansen does the same, though he also observes that there must be some contact points between these different time conceptions. Otherwise, we could not talk about a common phenomenon 'time'. Johansen finds these contact points in the existence of event and action time in modern society.[12] We cannot live human life simply as a cog in the great clock-time machine. We must also be able to follow our actions' concrete rhythm, for example, dropping lunch if we are not hungry or going to bed late if we are having an unusually good time. This sense for the unique dynamics of our actions is something we share with premodern man.

Is it entirely certain, however, that we do not also share the clock's universal, homogeneous time? It is clear that this time was discovered relatively late in history. But does that mean it was not there all along? The question is perhaps whether it makes sense to completely abstract from what we ourselves, today, consider truths, including about the world we ourselves have not inhabited. When we describe clockless societies, we describe an assortment of time series in light of our own time experience, and in archaeology, physics and cosmology, events are described as being ordered simultaneously or before and after each other in a bygone, but homogeneous time. This time is viewed in relation to today's clocks and we cannot forget or 'unlearn' the modern time concept. As such, it is also impossible to think that something is true without it. Must we, therefore, conclude that premodern man had his worldviews and we have ours, and, as such, we should not consider our own perceptions to be accurate conceptions of the world?

Physicists and other natural scientists typically answer these questions in the negative. Nature does not change throughout history, they will say. Indeed, from the Big Bang on, the universe has been governed by the same constants. Today we know more about these natural laws than did people at any previous time, something we have utilized on a practical scale all the way down to the nanotechnical and quantum-physical levels. Meticulous, detailed knowledge of physical time series enters the technical control of our surroundings, and because knowledge is based on universal, natural laws, that also makes it possible to understand the order and velocity by which things happened in an arbitrarily distant past. All knowledge of this type is, of course, based on modern clock mechanisms. But a physicist will not therefore conclude that past time sequences are something we invent rather than discover. This also applies to all that has taken place in premodern man's natural surroundings.

In particular, this idea applies to the simultaneity of various time sequences. As the story of life on the island illustrates, the simultaneity concept has subsidiary meaning in a clockless time experience. In modern clock life, by contrast, everything is synchronized according to a common standard. Yet only a couple of centuries ago, synchronizations were largely localized, with clock settings varying from village to village. But the formation of nation states was characterized not just by common measurements and currency, but by the country's inhabitants living by a standard clock setting. This idea requires having clocks that all show the same time, and the technical issues that arose from this synchronization process were an important inspiration for the man who revolutionized the modern physics time concept.

TWO

PHYSICS TIME: EINSTEIN

On 20 July 1969 Neil Armstrong set foot on the Moon with the famous words, 'One small step for man, one giant leap for mankind.' People the world over sat before their television screen, experiencing this historical event simultaneously as it occurred. 'Now it's happening,' people cried to each other in countless homes, 'Come and see!' Television signals, of course, require a split second to reach us from the Moon, so to be exact we would have had to watch Armstrong's foot touch the surface right before we viewed it on screen. This kind of precision, as we will soon see, is unavoidable in relativistic physics. But the delay was too slight to play any role in the viewers' experience of watching the Moon landing as it occurred. As such, the event was a concentrated expression of our modern awareness of existing in a single, universal 'now'.

In the previous chapters, we saw that people living under clockless conditions have only vague conceptions of existing in a similar simultaneity. Their time consciousness is linked to limited, concrete activities, and one does not always operate with a clear distinction between something being distant in space and distant in time. Following the European Renaissance, however, the natural sciences paved the way for the physical world to be regarded as a clockwork. Later, people became used to synchronizing their activities, first locally, then nationally and then globally. In keeping with this practical synchronization, the awareness arose that simultaneity was not limited to the global level: events occurring in the physical cosmos's most distant regions take place more or less simultaneously with what is happening on Earth. The consciousness of simultaneity becomes universal.

The great scientist Sir Isaac Newton (1642–1727) also operated with a universal simultaneity concept. Space is, according to Newton, an endless, extended container and all events in time occur simultaneously with other events at other places in space. Therefore, it is not just that states supersede each other locally, where I am right now, for example – first a sharp, then a rich and finally a round taste from a sip of wine. No, all that occurs 'here' occurs simultaneously 'there', with other events at other places in space.

This view of the relationship between time and space can be illustrated by a simple diagram where we imagine the past extending to the present along a timeline, and where everything happening in space at a single point in time is depicted by horizontal simultaneity lines. In the following time segment, two events are marked along this line. Even as person A is born somewhere on the globe, person B dies somewhere else.

Future

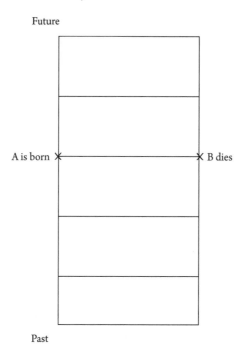

A is born ✕━━━━━━━━━━━━━━━✕ B dies

Past

The diagram shows that you can imagine time's passage in salami form, with space-slices cut across the timeline. For Newton, it was important that every event be located on one and only one space-slice, so there were no other possible ways of cutting the time salami. This idea corresponds to our notion that things either occur simultaneously or they do not. Either Armstrong set foot on the Moon at the exact same moment the candle on my television extinguished or he did not.

It was on this point that Albert Einstein (1879–1955) broke with Newton. In the universe, Einstein argued, unambiguous simultaneity does not exist and events that happen simultaneously for one person can happen in sequence for another. This idea creates another 'salami model' for time and space, without unambiguous space-coordinates for each physical event. Einstein's time salami can be sliced in many ways, with axes that cross each other.

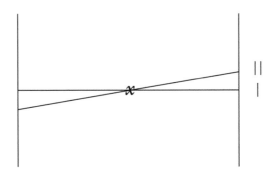

If the salami can be sliced in many ways, then there is no universal 'now'. If x represents the extinguished candle, then I, from my place on the sofa, would say it extinguished with Armstrong's first contact with the Moon's surface (I). However, if we imagine a person passing my house at great velocity, then he or she might justifiably say that the candle extinguished with Armstrong's second contact (II). The time salami, that is, can also be sliced diagonally, as the diagram shows. Simultaneity, therefore, becomes relative, something that can vary from person to person. This is the result of Einstein's relativity theory, which breaks with Newtonian physics and with our intuitive notions of a single, universal 'now'.

Before we take a closer look at the theory's rationale, we can remark that it was first presented in a short article from 1905, while Einstein was still employed at Bern's patent office. Not only did he have years of theoretical immersion in electrodynamics behind him, his day job afforded him extensive practical experience with clock synchronization, which, with the conversion to a new century, represented a notable technical challenge.

The mechanical clock disciplines and synchronizes human life, something that presupposes that clocks tick in conjunction and show the same time at the same point. Variations certainly arise, such as the current time zone changes between different countries, but keeping track of these time differences poses no great difficulty. The matter was more complex, however, before the institution of national clock synchronization. Since clocks were merely set according to the Sun, their time varied from city to city and from village to village. So long as people spent most of their lives within a local community, without significant contact with the outside world, this fact did not present much of a problem. With the development of the railway and other transportation and communication means, however, this system increasingly became impractical. Across the vast United States, there was typically one time for the railway company's schedules, one for the individual passengers and one for every station.

In today's society, the regulation of railways, air travel, schools, work, opening hours, radio and television programmes, the police and the military could hardly function without common time measurements, and national clock synchronization is a patent expression of the central government's ability to ensure its citizens march in step. Throughout the 1800s, therefore, measures for such synchronization were a high priority in European countries. This first occurred in cities like London, Paris, Geneva and Berlin where, through signals from a master clock, clock towers and watchmakers implemented a uniform city-time. In Paris hydraulic synchronization was attempted, among other things, with air pumped through pipes from the city's centre to the outskirts. But the telegraph's electric signals proved superior, as they finally did on a national scale as well.

The challenge, therefore, was issued to aspiring inventors to construct gadgets capable of uniform, effective clock synchronization. Through his work in Bern's patent office, Einstein met a number of such inventors, so he was well acquainted with the following technical conundrum: how best to synchronize clocks using electromagnetic signals, so we know precisely that an event in one place is occurring simultaneously with an event happening in another? Fascinatingly enough, it was the theoretical version of this same problem that resulted in Einstein's new time concept.

Galileo's Relativity Principle

We have seen how the idea of a single, uniform time is tied to modern clockwork. Without a device that 'runs like clockwork', time does not seem homogenous and universally quantifiable. We have also seen that physics regards natural laws as homogeneous and universal. Whether describing an event that happened right after the Big Bang, or inside distant galaxies, or back in the Stone Age, or among indigenous peoples in the Amazon, or in our very own kitchen, everything takes place according to the same natural laws, with the same explanations of how things change over time. This means our ideas of a universal time are not only connected to clocks as timekeepers. They are also tied to the universal physics that replaced ancient and medieval worldviews. Einstein represents a zenith in the development of this universal physics, which consequently treats all world phenomena equally.

For ancient and medieval peoples, the case was different. Things near to hand, things existing 'beneath the Moon', were thought to be composed of ordinary, earthly material and subject to ordinary, earthly laws. The stars up in the heavens, by contrast, were literally created of a lustrous, supernatural material that was subject to different, supernatural laws. Nonetheless, Earth was the natural resting point, in relation to which bodies must be provided with physical forces to move. In the Renaissance Galileo Galilei (1564–1642) and other pioneers broke with these ideas for the sake of a new science. Certainly, the Earth exhibits significant local differences; a man on the Moon will perceive entirely different things from a person in the Maldives. Yet, physics is not concerned with such differences. Through experiments, it uncovers fundamental natural laws that are applicable independent of time and place, and of movement and rest.

Even as Nicolaus Copernicus (1473–1543) showed that Earth is not some natural midpoint in the universe, so Galileo demonstrated it is no natural resting point. There is no natural resting point, he argued, for movement happens in relation to a body of reference, and what constitutes that body of reference is, in principle, arbitrary. Standing in a grove and watching a passing train, it seems normal to say that we are standing still and the train is moving. Movement is relative, however, and the train's passengers can claim just as legitimately that it is they who are still and we who are moving in relation to them.

The train, for its part, is tiny compared to the rest of the globe, so we tend to think of the Earth – or the grove, in this case – as 'fixed'. But it takes no great effort to realize that the Earth, on its wanderings throughout the cosmos, does not represent any absolute state of rest. In the name of fairness, so to speak, we realize that the train, too, can claim to be at rest. The physical reality here is that natural laws do not distinguish, and if I toss a ball along the train's corridor, it takes the same effort to give it a certain speed as it would on the ground outside. And if I pour myself a cup of coffee, I hold the pot directly over the cup, not a bit to the side to compensate for the train's movement. Galileo provides a picturesque description of these relationships, only replacing the train with a boat:

> Shut yourself up with some friend in the main cabin below decks on some large ship, and have with you there some flies, butterflies, and other small flying animals. Have a large bowl of water with some fish in it; hang up a bottle that empties drop by drop into the narrow-mouthed vessel beneath it. With the ship standing still, observe carefully how the little animals fly with equal speed to all sides of the cabin. The fish swim indifferently in all directions; the drops fall into the vessel beneath; and, in throwing something to your friend, you need throw no more strongly in one direction than another, the distances being equal; jumping with your feet together, you pass equal spaces in every direction. When you have observed all these things carefully ... have the ship proceed with any speed you like, so long as the motion is uniform and not fluctuating this way and that. You will discover not the least change in all the effects named, nor could you tell from any of them whether the ship was moving or standing still.[1]

Or let us take a more relevant, maritime example. On a cruise ship, there is not only a fitness centre, a swimming pool, a cinema and other comforts, but a physics lab. Additionally, there is a deck fountain that spurts water regularly and attractively in all directions. The world's leading physicists happen to be holding a conference on this ship, which means frequent visits to the lab. As long as the ship is moving nicely and evenly in calm waters, the physicists have no reason not to have confidence in the lab experiments. The restaurant's waiters pour coffee as they would on land, and the experiments function exactly as they would on land. In both cases, the reason is the same: all bodies move in relation to other bodies, and, physically speaking, none of the bodies are in any distinctive position. Therefore, we can say that the ship is as decent a body of reference for the physics experiments as the globe's fixed hills and dales.

This thinking flows into what is often called Galileo's relativity principle, something that forms an important background for Einstein's relativity theory. The fact that neither waiters nor experimentalists notice any difference shows that natural laws are neutral in relation to the choice of reference body. That is, the same physical forces are applicable everywhere – both on land and on a boat moving at 20 kph relative to the shoreline. These forces do not create the same speed relative to different bodies of reference, but they create the same change in speed. Perhaps we mobilize all our strength to cast a heavy ball onboard the boat so that it accelerates from 0 to 10 kph along its movement path. In relation to an observer on the shore, the ball's speed changes from 20 to 30 kph. The speed change in both cases is the same: 10 kph. Such alterations are explained by physical forces, and it is those forces that physicists interrogate in their labs.

Even as we formulate the relativity principle in this way, however, we must modify it. Namely, light and its propagation creates problems. Light spreads, as we shall see, at the same velocity in all directions, independent of what it is measured against. Like all physical forces, light speed is also independent of the reference body, which, according to Einstein, means light speed is a natural law.

This idea departs entirely from our daily experiences, and it is useful here to stick with boats. In our example, the ball rolled 10 kph relative to the boat and 30 kph relative to the shore. We could also have observed the water fountain on deck. Onboard the boat, which is moving at 20 kph, the water spreads at 10 kph in all directions. Relative to the shore, that means that the water jets spray forward at 30 kph and backwards at 10 kph along the path of movement. If we discount air resistance and other disruptive factors, we have no reason to doubt that a water fountain would behave approximately like that. A light fountain, however, would behave differently.

The Light Fountain

A few years ago, the Norwegian princess Märtha Louise was recognized for her glowing happiness, as well as the statement that she felt like a fountain of light. Let us imagine her on deck now, at the centre of a beam of light. Since a light fountain is analogous to a water fountain, we do not, strictly speaking, require flesh-and-blood royalty in the mix. An ordinary lamp or some other source that, akin to the water fountain, spreads light in all directions would function as well. In any case, what happens with a light fountain is astonishing:

(1) A person on land will observe that light propagates at a speed of 300,000 kph in all directions. Contrary to

what happened with the water fountain, the boat's path of movement makes absolutely no difference.

One might think there is a simple explanation for this: that light propagates in a 'light medium' like sea waves in water. If someone on the boat throws an object into the water, waves form that move in all directions at the same speed relative to the shore and not to the boat. (If the boat is moving fast enough, it will even draw the waves along its path of movement.) Before Einstein, the dominant viewpoint among physicists, therefore, was that light propagated through an invisible light medium, a so-called ether. But there was one thing this ether theory had particular problems explaining:

(2) In contrast to water waves, light waves also move at the same speed relative to the boat.

The result of (1) and (2) is truly strange. Both in relation to the boat and in relation to the shore, light disseminates at the same speed. It is a simple fact and Einstein was not alone in affirming it. And it is a fact worth acknowledging, for example, through comparing the light fountain to the water fountain on deck. For a person onboard the boat, water spreads at the same speed in all directions. Over a certain period of time, all drops will equally distance themselves from the individual, so that she or he will exist in the water fountain's centre. Relative to a person on the shore, however, the water jet moves forward at 30 kph and backwards at 10 kph along the path of movement. Over a certain period of time, the entire fountain will distance itself from the individual, who cannot then be said to exist at its centre at all.

Now take the light fountain again. Both relative to the boat and relative to the shore, the beams spread at the same speed in all directions. Relative to the shore and relative to the boat, therefore, they cover the same distance at the same time. That means that even if the boat is moving relative to the shore, both the boat and a person onshore stand at the light fountain's centre. In Märtha Louise's case, that almost entails a democratization of the fountain. Not only do I exist at the centre of her radiance when I stand there talking to her, but as I go on my way, the light transmitted surrounds my movements as much as hers.

The fact that concrete, physical light behaves in this way was not, as mentioned above, Einstein's discovery. But now we can at last understand the time salami from pages 45 and 46. If we are to explain the light fountain's behaviour, according to Einstein, we must slice the salami crosswise. In order to illustrate this idea, we will begin by depicting the light fountain with two lines drawn diagonally across the time and space diagram.

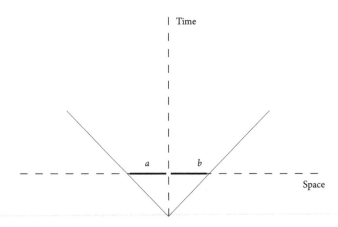

We can recall that the vertical line represents time's movement from past to future. The horizontal line represents space, which intersects the timeline perpendicularly at a point. That is, the points along the space line are simultaneous. The diagonal lines represent the light fountain, or lightcone, as physicists say. The further we advance up the timeline, the further the light moves right and left into space. The fact that the diagonal lines appear the same in both directions signifies that the movement occurs at the same speed in all directions. A person standing still in space will simply 'shift' along the timeline. As the figure shows, that person is at the lightcone's centre the entire time (distance a = distance b).

Also, a person moving relative to the first individual can be represented by another diagonal line. This occurs in the diagram below, where a new timeline denotes one person changing location relative to the other, who stands still along the original, dotted timeline. Yet, how can we graphically depict that the person moving along the new timeline also exists at the lightcone's centre? How can the spatial distance of the individuals to the beam be equal in all directions? That problem seems insoluble. We can solve it, however, if we tilt the space line as we have already tilted the timeline. Then the segments c and d are equal.

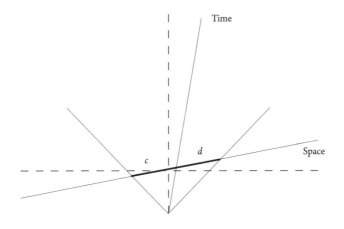

If we grasp the connection between the diagram and the light fountain's behaviour, then we grasp something essential about Einstein's theory. Since Märtha Louise is standing still, she is located along the original, dotted timeline, and along the original, dotted space line she exists at the light fountain's centre. If I move relative to her, then I am located along the new, diagonal timeline, and along the new, diagonal space line I exist at the light fountain's centre. We are both at the centre, simply located on our individual space lines.[2] And each space line represents everything that is occurring simultaneously. Both Märtha Louise and I can be at the light fountain's centre as long as what is simultaneous for her is not simultaneous for me. In this way, Einstein's time salami is explained.

A single 'now' contains a single, simultaneous slice of space. On that point, Einstein and Newton agree. But for Einstein it is not once and for all determined what each simultaneous space-slice contains. For a person on the original timeline, it contains all that exists on the original space line. For someone moving along the new, diagonal timeline, it encompasses all that exists on the new, diagonal space line. Because I was sitting at home in the living room, the candle extinguished with Armstrong's first contact with the Moon's surface. For someone driving by, it extinguished with the second contact. Both claims are true. And not only does it appear that way, it *is* that way, if we are to trust Albert Einstein. Otherwise, two people who, at a given time, find themselves in the lightcone's centre cannot remain in that same lightcone's centre when they move relative to each other. Yet, that is exactly what they do.

In other words, we must append the phrase 'relative to what' if we wonder about the simultaneity of two events. The same is true if we question how long something lasts. As many people today know, Einstein's theory leads to the conclusion that things take longer if they move relative to a reference body than if they do not. A twin who shoots into space aboard a spaceship is slightly younger than his previously same-aged brother when he returns home. The relativistic explanation for this is that, when measured relative to things occurring on earth, the ageing process takes longer than when measured relative to other processes onboard the spaceship.

As such, time is 'relative' in relation to who is measuring it. That might sound strange, perhaps because we are used to assuming that there is a specific duration in everything that happens, independent of other things occurring. In physics, however, we must relate time series to other time series in order to quantitatively determine them.[3] From this perspective, it is not so odd to say that something has one duration relative to one measuring system and another duration relative to a different measuring system. It is just as natural to say that a boat has one speed relative to the shore and another relative to another boat. Conversely, asking how long something 'actually' took is as meaningless as asking how quickly something 'actually' travels, independent of other things.

The relativistic aspect of Einstein's theory assumes that he does not regard the duration of event sequences as a quality inherent to what is happening, but rather as correlations, relationships between things that happen. Time is relative because it is relational, something that can simultaneously disclose the limits of a physical theory regarding human life time.

So What About 'the Present'?

At the beginning of the chapter, I mentioned the well-known experience of finding oneself in a universal 'now': 'Crack the champagne … *now* he's setting foot on the Moon!' Still, everything we do 'now' happens within a past and future horizon. Perhaps I know I have to be up early tomorrow in order to catch a flight to New York, where I will be spending the next two months. Before I go to sleep, though, I will crack open the champagne, which I recall buying the previous week and taking from the fridge a few minutes ago. As a result, even if my entire attention is focused on the present, it occurs in light of the near and far past and future. Conversely, the consciousness of past and future always merges with the consciousness of being present in what we are doing 'now, right now'.

But we have already seen that this 'now' is not as straightforward as we first believed, if we consider it to be common simultaneously to all people. Things that are simultaneous for one person need not be simultaneous for another. That does not imply that we should stop talking about simultaneity altogether, just that, strictly speaking, we must remember that simultaneity is not 'absolute'. Yet we have also seen something else: simultaneity is relative because it is relational, something that poses an important difference between the concepts of 'presently' and 'simultaneously'. In physics, one does not speak of past, future and present. One simply states that what is occurring is related to other things that happened before, after or simultaneously with it. That does not mean the same thing as something having occurred in the past, future or present.

For example, maybe I know the champagne will be opened at the same time as the Moon landing at 9 p.m., although I do not know whether that event is happening 'now'. Even though I know that something is occurring simultaneously with a specific physical event, I do not automatically know that that event is now. Furthermore, I can be aware that my great-uncle is turning seventy a year before my great-aunt, without also knowing whether both birthdays have occurred, whether neither of them has occurred or whether just one but not the other has occurred. Knowing one event happens before another is not the same as knowing whether that event is past, present or future. Finally, I can know that a film lasts longer than another film without knowing the precise duration of each. Recognizing the relationship between two time sequences is not the same as knowing how long they last.

All of this demonstrates that there are clear distinctions between physics time and people's experienced time, something that perhaps becomes even more obvious in our connection to our emotional life. Imagine a football fan who shouts, 'Now the match is over!' (with 3–0 to Liverpool) . . . but does not express the same exact thing with, 'The match ended at 10 p.m.' In physics, it plays no role whether 10 p.m. indicates a point in the past, future or present. But the experience of an event as present is completely different than the experience of that same event as either past or future. Or, take the way we are oriented towards past and future. I can look forward to a future event and I can bitterly regret a past happening. But it is senseless to look forward to something I have done or to regret something I will do.

Einstein was well aware of this distinction between physics time and human life time. Indeed, as the conversation he had with the philosopher Rudolf Carnap (1891–1970) shows, Einstein regarded this difference as a principle limitation to physics' contribution to the knowledge of human life:

Once Einstein said that the problem of the Now worried him seriously. He explained that the experience of the Now means something special for man, something essentially different from the past and the future, but that this important difference does not and cannot occur within physics. That this experience cannot be grasped by science seemed to him a matter of painful but inevitable resignation. I [Carnap] remarked that all that occurs objectively can be described in science; on the one hand the temporal sequence of events is described in physics; and, on the other hand, the peculiarities of man's experiences with respect to time, including his different attitude towards past, present, and future, can be described and (in principle) explained in psychology. But Einstein thought that these scientific descriptions cannot possibly satisfy our human needs; that there is something essential about the Now which is just outside the realm of science.[4]

The most important reason that the concept of 'now' does not enter into physics is that it designates a subjective standpoint in time. It applies to the position that I and other people happen to inhabit right *now*, and this does not interest physicists, who do not describe the world from any specific human perspective.

The fact that things occur before, after or simultaneously with each other in an objective time sequence is something we can verify independently of specific human perspectives. Therefore, it is uninteresting that to me my grandparents lived in the past and my grandchildren will live in the future. It is enough to determine that certain people in the twentieth century live for such-and-such length of time before their descendants two generations later. This fact will always be true, regardless of whether my grandparents live in the past (as they do for me) or in the future (as they did for my great-grandparents).

Einstein himself was well aware that physics' objective series of events results in a 'timeless time' where my individual death, as an essentially future occurrence, is not distinct from my individual birth, as an essentially past occurrence. Indeed, as he said in order to comfort the grieving relatives when his close friend Michele Besso passed away: 'And now he has preceded me briefly in bidding farewell to this strange world. For us believing physicists, the distinction between past, present, and future is only an illusion, even if a stubborn one.'[5]

There is a clear difference between physics time and human, experienced time. We can believe, of course, that these two time concepts are connected in some way, that we are not, in fact, talking about two distinct times. This is a question to which I will return. Meanwhile, since we have turned our attention to the twentieth century's most important 'time physicist', let us now take a look at that same century's most important time philosophers.

THREE

PHILOSOPHY'S TIME: BERGSON, HUSSERL, HEIDEGGER

Wherever you turn, you find movement. Turning is itself movement, and if you glance out of the window, you might see falling raindrops, a bird flying by or a leaf blown by the wind. Even when lying still on a sofa reading, your chest moves with your breathing and your lashes move when you blink. I am almost tempted to say that movement is one of the most plentiful things on earth. And like every change, movement takes time. Movement, moreover, is often employed to provide a picture of time, when we refer to, for example, 'time's arrow', 'time's flight' or 'time's passage'. Without movement, it is difficult to illustrate time; without time there is no movement. Movement and time are intimately connected.

That is not to say, however, that other forms of change are less important. Still, one characteristic of change, as I discuss in Chapter Five, is especially evident with movement: change does not exist as a series of separate events, but because one and the same thing shifts position. If, instead of a bird flying by, you see the bulbs in a lighting display successively flashing, one after the other, such light signals would constitute a series of events – that is, things that happen. Seeing movement, however, does not mean watching different events occur at different positions, but observing a single thing as it shifts position. So even if the light signals follow each other from left to right, that does not constitute movement from left to right. That means that to many, the movement phenomenon fares 'poorly' in relativistic physics.

In Einstein's special relativity theory, everything revolves, as we have seen, around light's propagation, and time and space are regarded as ways of classifying separate light signals. If, for example, a signal is transmitted, reflected and received again, it is not a matter of a single, enduring object moving from place to place, but of three signals, one for each event (transmission, reflection, reception). Just as objects placed side-by-side in space represent different pieces of the world, so every signal forms an individual, independent piece of reality. We do not have one single thing shifting position, but three different events that do not shift position. As with the lighting display, no actual movement has taken place.

Now, no one will deny that we human beings experience a world of movement and change. But in this discussion of the relativity theory, the physical reality behind this experience is made up of isolated events. Because we nonetheless have subjective impressions of movement and change, the process can be compared to the images that make up a film. Presented sequentially on a canvas, they create the illusion of 'flow' or movement:

> There is no dynamics within space-time itself: nothing ever moves therein; nothing happens; nothing changes ... Imagine a film has been taken of what occurs in the world, that this film has been cut into its individual frames, and that these have been stacked on top of each other. The

result is similar to space-time ... In particular, one does not think of particles as 'moving through' space-time, or as 'following along' their world-lines. Rather, particles are just 'in' space-time, once and for all.[1]

Events within the theory of relativity exist independent of each other in a kind of timescale, as a parallel to objects placed at different points in spatial landscapes.[2] Like articles of space, so articles of such time cannot exist within each other, as they do, in a certain sense, when we see something in motion: in the present, the subsequent position appears as something future and the prior position as something past. We experience, that is, not just a series of nows, but the way in which past and future make themselves relevant in the present. This is what gives us the feeling of 'time's flow', and without this 'flow' we must stop and ask if, in the realm of relativistic physics, time's true nature has been uncovered, or if it has rather been overlooked.

We will now turn our attention to three philosophers who agree with the latter idea: Henri Bergson, Edmund Husserl and Martin Heidegger. They do not discuss relativistic physics to any great extent, and they are less concerned with movement than with the more profound changes in human life. More generally, however, they claim that as long as the sciences dismiss the future's and the past's presence in the now, that which makes time time is lost.

Henri Bergson (1859–1941)

Henri Bergson was born in Paris to an English-Jewish mother and a Polish-Jewish father. After Bergson won a prestigious

mathematics prize, his maths teacher complained about his leav-
ing the natural sciences for the humanities: 'You could have been
a mathematician; now you're nothing more than a philosopher.'
But a philosopher he became, with a series of influential works
on time, consciousness, development and life. A recurring theme
in his work is that everything in the world is in time, that time is a
form of consciousness, and that consciousness, therefore, forms
the essence of existence. His works are sharply conceived and
concisely written with lines of reasoning that are not always easy
to follow, but which are always accompanied by precise observations. That contributed no doubt to his distinction and fame, and
he eventually achieved institutional status in French intellectual
life. For his philosophical works, he was awarded the 1928 Nobel
Prize – in literature!

The question of how past and future can exist in the present takes us directly into the heart of Bergson's time philosophy.
That is, Bergson was particularly focused on the way the past
remains in the present. In order to understand this idea, Bergson
developed his own special concept of *la durée* – duration.

Quantitatively, the past does not remain in the present, just
as one part of space is not present in another, equally sized portion of space. Take, for example, six lines located next to each
other on a piece of paper. The lines have an obvious number and
there is a specific numerical relationship between the leftmost
two and the four others. The distinction between objects that
exist outside each other in space is clear enough that counting,
addition and other mathematical operations yield unambiguous results. Such is also the case with spatial representation of
time points on a calendar or clock face. None of these charac-
ters is contained in another and the sum of multiple time points

is just as unambiguous as the sum of the six lines on the paper. Considered in this way, the past cannot remain in the present.

According to Bergson, however, time is not as quantitatively divided as space. To the contrary, as long as we focus on the clock's or calendar's spatialization of time, we lose that which is time-ly: that which makes time time. For if an element of time, the past, remains in another element, the present, it does not occur in the same way as the number four is part of the number six, as the sum of two and four. Time's components are part of each other in a more dramatic way, as would be the case with space if the two leftmost lines were part of the two rightmost. When it comes to space, they are not, so long as we are talking about a specific number of lines. So how is this kind of 'presence' possible in time?

The special thing about time is that its components are qualitatively, not quantitatively, present in each other. Not only that, if we regard time as independent of space, we see that it does not actually work to divide it quantitatively. This is perhaps not as strange as it first sounds, for it is the same with that which we humans are most familiar: consciousness. Consciousness first makes time comprehensible, Bergson argues, for time's elements are 'put together' just like those of consciousness. Consciousness cannot be quantified, nor does it consist of separate parts. Let us first take a look at this missing quantification.

'Five stars', 'three stars' – as newspaper readers, we are accustomed to having the quality of films, books and musical works graded. That is all well and good; it informs us that the reviewer considers something average (three stars), excellent (four stars) or outstanding (five stars). Or do these grades make the assessment even more precise? It might seem that way, but if we pause to consider, we realize that is not the case, and Bergson can help us understand why.

We humans live a life filled with moods, emotions and other experiences. Some of these conditions bring sorrow, others joy. Some are strong, others weak. Some make a lasting impression, others only graze our minds and are forgotten as quickly as they appeared. Comparisons like these help make some experiences seem important, others less so. But one type of comparison seems misplaced: quantification. We do not ask, 'Was the first joke three or four times as funny as the second?', or 'Do you agree that the mood of the second painting was 50 per cent more melancholy than the first?' Although one sorrow can seem much worse than another, and one joy much greater than another, the difference itself cannot be quantified.

When emotions do nonetheless get compared, it is done qualitatively. If we say that one joy is 'greater' than another, perhaps that is because it feels richer, truer or more human. Even if we speak of 'more or less' genuine feelings, the difference between a true joy and an illusory or false one is qualitative, not quantitative. The quality there might well be tied to significant amounts of something or other, like a rich joy rooted in major life experiences. But it does not consist of a sum total of joy atoms.

In the same way, I can contribute a personal example. Many will insist that a wild strawberry tastes better than a garden strawberry. I will nonetheless claim that a large garden strawberry tastes better than a small wild one because the sweet mouthful gives an experience of fullness that beats the wild strawberry. Size plays a role, but this difference is qualitative and not quantitative. Even though the garden strawberry tastes better because it is seventeen times as large as the wild strawberry, it is senseless to say that it tastes seventeen times as nice.

The reason, Bergson would say, is that when we compare experiences, we do not compare amounts of a single, homogeneous quality, as when we compare stronger and weaker notes. The difference is more like the contrast in sound produced by an isolated string and an entire symphony orchestra.

But can we at least quantify the intensity of taste and other physical sensations? Bergson's view is that we cannot, not even with something as simple as a needle prick. We first experience it as a light brush against the skin, which then transforms into a tickle, then an itch, then discomfort, then pain, then worse pain, and that which makes the pain steadily worse is not degrees of a single, simple quality but the fact that we tense the muscles over steadily greater portions of our body to withstand the pain. Just as the garden strawberry is qualitatively better than the wild strawberry due to its size, the pain is qualitatively worse the more parts of the personality it affects. In both cases, it is a matter of different qualities, not degrees of intensity of a shared, homogeneous quality.

The comparison using the symphony orchestra also shows that the relationship between whole and part is different than that among an assembly of homogeneous quantities. Just as the orchestra is composed of multiple instruments, so multiple elements enter into positive or negative experiences. On the one hand, we never have a simple, undifferentiated holistic experience, just as there is no orchestra without individual instruments. But, on the other hand, the whole characterizes the parts in a distinctive way. The sound of an oboe is even more beautiful against other more sonorous instruments, and the taste of the larger strawberry serves to characterize the taste elements it shares with the wild strawberry. What distinguishes the experience is the way in which the whole and the parts inform each other. The case is even clearer with moral emotions, Bergson argues, than with straightforward sensory experiences.

Take, for example, compassion. At its root is a feeling of suffering when others suffer. Typically, we do not wish to suffer, so a natural response would be to turn away from one's fellow man in order to quell the emotion. On the other hand, we can engage in others' suffering for our own self-interest, because we know that next time we might be the one who needs help. True compassion, however, first emerges when we in all humility realize that our own interests do not count more than those of others. Ultimately, this reduction in our own significance can result in heightening our self-esteem. In true moral compassion, all these elements are generally present. It is not as though we have fear in one corner of our consciousness and humility in another, as we have some lines at one place on the paper and others at another place. All are generally present.

Still, what is the relevance of this analysis of consciousness to our analysis of time? Consciousness allows for something that is impossible within inanimate nature, namely, that one condition within the world permeates another. Compassion's elements permeate each other, the whole permeates the parts, and that is the way it is with time as well. Every step in time's development forms a new whole that permeates the next 'now', and if the past remains in the present, this happens, according to Bergson, in the same way as we experience a melody as an organic unity of its parts:

Might it not be said that, even if these notes succeed one another, yet we perceive them in one another, and that their totality may be compared to a living being whose parts, though distinct, permeate one another just because they are so closely connected?[3]

The past also permeates the future in a way that is only possible in consciousness, and this is Bergson's primary concern within philosophy: everything that exists is in eternal change, change only happens in time, time is consciousness, so, at its root, all that exists is consciousness. Consciousness, in turn, is inextricably tied to life, because, as he indicates in the above citation, the relationship of parts to the whole is like that of organs to the organism they serve. It is the whole that determines the function of each. In time, every new present experience shapes a new whole from present and past, and this whole ensures that the past, which permeates the present, is never exactly as it once was. This idea is familiar to anyone who has experienced the sweetness of that first encounter – and perhaps the irritation of the tenth encounter.

'The first time, the first time, it lends rank to many a small thing,' as the Norwegian author Henrik Wergeland put it. Who has not experienced the joy of that first love, of that first swimming stroke or of that first strawberry? And this special first-time joy cannot be repeated. In contrast, what we know only from repetitions is the irritation about the tenth trumpet scale from the room next door or the tenth bark from the neighbour's dog. Originally, the first bark was not irritating, but it eventually became so, precisely as it entered the past's permeation of the present.

This is true of everything that occurs within Bergson's notion of time. Despite the fact that all the links in a chain of events are physically identical, new time experiences are permeated by past time experiences, which they then help to change. And as it is with experience, so it is with time. Time is never itself the same, so it is not homogeneous, and it cannot be divided and quantified. Like time in clockless societies, it is qualitative, heterogeneous,[4] and an eternal source of new things under the Sun.

In contrast to Husserl, at whom we will now take a closer look, Bergson is not primarily concerned with the fact that consciousness seems to be outside itself when it relates to past and future. Also, in space, consciousness always contains more than we, at any point in time, perceive. If we live on the third floor, for example, we have a perpetual sense that there are two floors beneath us, even though we are not always thinking of them. Yet Bergson's thesis is more radical when it comes to time. The point is not that consciousness extends to distant events, but almost the opposite: that distant events are gathered in present time.[5]

Edmund Husserl (1859–1938)

Edmund Husserl was a profound German thinker who, in contrast to Bergson, would never have been a candidate for a Nobel Prize in Literature, even if he had kept going for centuries. For a Nobel Prize in Philosophy, though, he would have been an obvious candidate. Presumably, only Ludwig Wittgenstein might give him a run for his money as the twentieth century's most influential philosopher. Husserl's ideas, especially in the realm of European phenomenology and existentialism, have left lasting marks. Heidegger was Husserl's student; Maurice Merleau-Ponty (1908–1961) further developed his concept of the lifeworld; and Jean-Paul Sartre's (1905–1980) discussion of human consciousness would be unthinkable without Husserl's concept of intentionality. Let us now take a look at these two last ideas, which are also important to Husserl's time analysis.

'Intentionality', in everyday speech, means someone has a purpose ('What was your purpose/intention in doing that?'). And if we have a purpose with something, then our consciousness is directed towards that thing. For Husserl, intentionality meant that consciousness has just such *directedness* towards things, and he argued that this orientation comprises an essential characteristic of our mental lives: consciousness not only consists of experiences like 'red', 'painful' or 'sour', but is directed towards conditions in the world. We see this clearly in love, hate, hope, fear, relief . . . the experience would be impossible without loving, hating, hoping, fearing or feeling relief *about* something or someone. As such, we are also engaged in a human lifeworld, which encompasses much more than that which at any one time meets the eye.

For example, if I drive by a church, I see a facade of smooth brick. But that does not mean that I perceive only the front of the bricks. I also see the entire building placed into specific surroundings. And I also know that the building has a back, a solid foundation, a roof and an interior with countless details, which themselves have backs and interiors. By thinking I see a church, I mean all the above, not just my changing visual impressions. Most importantly, I see the church as a three-dimensional object extended in space – a space in which I know I can move around, so I could see the building from behind. It is true that I only perceive the church from my place along the road. But in thinking I see a church, I know I could observe the building from different positions. In such a way, I perceive the structure all at once, as if I saw it from another location than the one I actually occupy.

Some of this also applies to time. In space, I perceive not only what is directly present to me right here, and in time I perceive not only what is directly present to me right now. Our conscious perception of a time sequence does not, therefore, correspond to physics' series of isolated events. According to Husserl, it *cannot* do that, because if we perceived time in this way, we would not be conscious of any time series at all. No matter how slight the course of events that we register, if it is events in time that we are perceiving, then we are sensing not only something present, but something past and something future.

Like Bergson, Husserl often uses music as an illustration, and the basic insight is that when we experience a variety of sounds in a musical piece, two things are excluded. In the first place, we do not sense all the work's parts simultaneously. If we did, we would only have one giant chord, no melody and certainly no musical piece. And a piece of music is not composed of simultaneous parts, we must also comprehend a series or time sequence of tones. So what about something as exceptional as the opening chord to the Beatles' 'A Hard Day's Night'? Does that chord not amount to an independent piece of music? Well, we can say that it only achieves its musical effect in relation to the rest of the song.[6] Anyway, Husserl's point applies to melody and rhythm.

In the second place, to regard musical works as successions of isolated notes is as ludicrous as regarding them as harmonies. In melody and rhythm, subsequent and prior notes sound together in each part-experience. Otherwise, we would have nothing we could call melody or rhythm. Naturally, subsequent and prior notes do not resound in the same way as do present notes, or all we would have is a muddled chord. But that means that if the notes do not resound as present notes, they resound as absent ones, as Husserl's analysis of 'directedness' as a type of present absence indicates.

This idea applies not just to musical comprehension, but to time consciousness in all areas of life. We do not always set ourselves clear, conscious goals concerning the future. Yet, clear or unclear, conscious or unconscious, we always have some expectation regarding the near future. Therefore, it is always possible to surprise a person, something that would not be possible if they experienced the present in isolation, without future awareness. This future awareness, which always surrounds our present consciousness, is what Husserl calls 'protention', in contrast to direct present awareness, which he calls 'impression'. Every impression, furthermore, has a 'comet tail' of past consciousness that itself rests on earlier impressions. This type of past consciousness he calls 'retention'. Because this idea applies to every form of time consciousness, it must also encompass natural scientific observations.

Take a ball rolling down a ramp. Even though we only see the movement in a flash, it still consists of multiple time phases. Otherwise there would be no movement, just a photographic still-life, a snapshot. As soon as we see a movement, we see the ball travelling to a subsequent position from the prior one. All time phases are not, therefore, equally present, and in Husserl's terminology one phase has a status as retention, another as protention, in relation to an impression. At the same time, Husserl believes that such time consciousness poses a fundamental reality in the world as we perceive it, not only as we experience it subjectively in thoughts, ideas and fantasies. These notions emerge most clearly from what he has to say on the concept of retention.

Retentional past awareness is not the same thing as memory, for it is too present in time for that. To recall or remember something means to let present conceptions represent an absent past. As a reader, for example, you can put a book aside and dwell in memory on all the wonderful things that happened last summer. Remembered images stream past, images you experience now and that apply to things you perceived then. But Husserl's retention does not pertain to something we have perceived before, but forms an element in what we are perceiving now. When it comes to remembered images, this applies to the present succession of those images, not to what they represent from the past. Retention does not represent anything, according to Husserl; it presents to us a world of sensory events. It is the fundamental level that is entailed in mental and physical images of the past.

A video clip, for example, can reproduce happenings from last summer. But we must sense the movement in the images, and we sense it directly, not through images of the images. More generally, not all perceptions of time series can be indirect, products of perceptions or images that *re*present other perceptions. Some of these must also be direct. Before we can represent things in the world, something had to have presented itself to our senses. Remembering something, meanwhile, is not to perceive something, but is another way of relating to the world.

Do you 'remember' the beginning of this sentence? That seems like a strange thing to say. But no doubt the beginning was 'retentionally' present before you began the next sentence. The contents of this book's first chapter, by contrast, are not present in the same way. You can try to recall them, however, by shifting your attention away from the page you are now perceiving to what you perceived when you read it.

What all this means is that time consciousness is not only supplemental to that which occupies us in the world, as when we are 'absently' concerned with making plans or remembering the past. The matter is more fundamental than that, permeating everything we do as we do it, and placing its stamp on the entire human lifeworld. Indeed, according to Husserl, it forms a basic characteristic of every consciousness, so that the concepts of time and consciousness essentially bleed into each other: with his I-consciousness, man is time's source. Some of the background for this perspective can be found in Husserl's problematization of his own theory.

A troublesome point in Husserl's time philosophy is that it is difficult, in practice, to draw a clear distinction between retention and memory. I used the beginning of a sentence and the book's first chapter as examples. But even our remotest past can, in a way, be 'present' in what we do, without us consciously thinking back on it. We perceive the world in light of previous experiences, after all, not only in light of what we just experienced. If we continue with musical illustrations, it could be a matter of the contrast between the orchestra to which we now listen and the one we heard before. Such contrasts can colour every second of the actual music experience. So even though Husserl's distinction between memory and retention is in one way clear, it is not entirely clear how, in practice, we should keep these two concepts separate.

Here we find elements of Husserl's philosophy that point beyond the borders of individual time experiences and towards the relationship between different time experiences within the individual's life and within history. This underscores another problematic point in Husserl's theory.

If we perceive a succession of events in the world, it is always in relation to a conscious impression, retention and protention. But particularly if these phases are extended over a significant timespan, it seems clear that they succeed each other in time. Would not the same logic then apply to each? If we experience 'lunch at noon' first as protention, then as impression and then as retention,[7] should we not perceive this succession relative to a new protention, impression and retention? And what about this new protention, impression and retention – would not the same be true of them, and so on into infinity? If we think like this, there seemingly emerges an endless series of time consciousnesses, something that appears absurd. But how can we avoid this idea? In order to find whether this problem can be solved, we must take a closer look at how it arises.

Husserl's point was that a series of now-experiences is not sufficient in order to talk about time consciousness. For something to be perceived as a time sequence, elements of the future and the past must also exist in recent consciousness. As such, consciousness gathers time's various phases into a unity, and this becomes a problem if consciousness is no more uniform than the phases it supposedly combines. Then we can compare consciousness to a bunch of faulty excavators trying to move a sandpile. The excavators are no better constructed than the sand grains they are trying to move, and at each use they collapse into a heap of metal parts. If all the excavators are equally defective, there is no point bringing in new ones. Indeed, having an endless supply would not help in the least.

The excavators, therefore, are unable to form a unity from the sand assortment, and to the extent protention, impression and retention succeed each other in time, they cannot create any unity from consciousness's time assortment. This is also what Husserl is getting at when he indicates in places that true time consciousness does not consist of phases that develop over time. Rather, it is a structure of past, future and present that is not itself located in time, but that creates the experience of things occurring in time. As such, time consciousness is itself time's source. This idea perhaps sounds mysterious, but Maurice Merleau-Ponty, inspired by Husserl and Martin Heidegger, claims enthusiastically that we indeed must view the matter thus. Time consciousness must itself be timeless, he argues, or otherwise we are left with an endless series of consciousnesses.

Instead of saying that time's phases succeed each other, Merleau-Ponty suggests that we compare time to a river – not with the river's constantly shifting water but with its constant, immutable form. This changeless river would not exist without its wellspring in a human consciousness that unites past and future in the present:

> We can say that ultimate consciousness is 'timeless' (*zeitlos*), in the sense that it is not intro-temporal. 'In' my present . . . there is an ecstasy toward the future and toward the past that makes the dimensions of time appear, not as rivals, but as inseparable.[8]

The specific terminology here regarding time's 'ecstasy' is something Merleau-Ponty has taken from Heidegger's theory that actual time is the root of all change's changeless, 'ecstatic' structure. We will now take a closer look at this idea.

Martin Heidegger (1889–1976)

Heidegger was bórn in the tiny south German town of Messkirch. His upbringing was strictly Catholic, and after a few years attending a Jesuit school, he began to study theology before switching to philosophy. At the University of Freiburg, Heidegger was an assistant to Husserl, who made a lasting impression on his thinking. In 1923 he achieved a professorship at Marburg, and in 1928 he returned to Freiburg as a professor of philosophy. There he remained for the rest of his life, spending a good portion of his time also at his Black Forest cabin. Politically, Heidegger leaned towards Nazism, holding a membership in the Party from 1933 to 1945, something that makes the relationship between Heidegger's politics and philosophy a fraught theme. An important theme in Heidegger's Nazi sympathies was contempt for the modern industrial society, itself tied to a strong anti-Americanism.

In Heidegger's main work, *Being and Time*, we find a philosophical activism. The significant thing, according to Heidegger, is not whether a theory is true or false in a traditional sense, but instead that something happens; that we enter into our time for the sake of uncovering new life forms and spaces for action. There is, of course, a certain connection between Heidegger's political attitudes and this philosophical activism. The Nazis, despite everything, ensured that things 'happened' within the oh-so-decadent Germany. Still, Heidegger's philosophical position can also be evaluated independent of the political circumstances, and Merleau-Ponty's reference to Heidegger's time analysis is a suitable departure point for this consideration.

We are almost driven towards this analysis as an alternative to sequential time, that is, time understood as a series or sequence of events. If present consciousness encompasses the past and the future, then the present becomes almost too narrow to contain it, meaning such consciousness cannot be located in a specific event. That is precisely what Merleau-Ponty intends with the word 'ecstasy', which comes from the Greek *ek-stasis* or 'to stand out'. The point is that we, through our time consciousness, stand 'outside of ourselves', placed though we are in the present. The full truth surrounding us humans includes a past and future time horizon that cannot be confined to specific steps within time's sequential unfolding. As Merleau-Ponty indicates, it is not so difficult to understand this past and future time horizon as immutable.

It is always now. No matter where in time we happen to be, we can always refer to the current point of time as 'now'. Things change in time, but the fact that they exist 'now' never changes. The same is true of past and future. All time points are past and future when viewed relative to some other time point, an idea familiar to all of us. Heidegger's specific twist is to give this formal truth an existential meaning. A life within a time-less, ecstatic consciousness, wherein every time point is present, past and future, is different from a life simply incorporated into a chain of events. Every time we 'arrive', we simply encounter a new departure point from which to set new goals we can then work towards 'arriving' at.

This gives time a strange double character. On the one hand, we can set ourselves concrete goals that we then try to achieve at a future present. A specific element in the time sequence is thereby transformed from possibility to reality. On the other hand, we bear past and future with us as an open, indefinite horizon, and it is man's fate never to cross that finish line. As such, the time horizon becomes an eternal source of frustration, something no one has perhaps described so fervently as Samuel Beckett: life consists in waiting for Godot, and Godot can never arrive, because then you no longer have a life that consists of waiting for Godot.

Like Bergson and Husserl, Heidegger also views time relative to our human consciousness as the unity of past, future and present. In contrast to Bergson, however, he does not devote much attention to the purely mental time experience. Instead, he further develops Husserl's concept of the lifeworld as the frame that surrounds our comings and goings. Man, according to Heidegger, is primarily a being that exists in the world through practical engagement with his surroundings. What separates us from other physical things is not that we are an immaterial consciousness or a soul without a body. It is that we are present among other things in a way that differs from other things. Rocks and trees do not prevail in the world in such a way that their existence seems to 'concern' them. But our human existence 'concerns' us, since we experience existence as more or less meaningful. If we speculate on how that happens, we will uncover aspects of human life that not even Husserl perceived.

For Husserl, mankind's consciousness is directed intentionally towards conditions and objects in the world, as when we love, hate, fear or desire something. Consciousness, therefore, is goal-oriented, and goals are things to which we can freely and rationally relate. Heidegger agrees with this idea, but he adds that every more or less rational choice occurs against a backdrop we do not choose, namely, the mood in which we happen to find ourselves. When we feel light and easy of mind, we tend to make different choices than when we feel dark and depressed, and such moods are not things we elect to be in. We can certainly try to influence our state of mind, but that, in turn, happens against the backdrop of another mood. According to Heidegger, then, all human life choices occur against the backdrop of an underlying mood we did not select.

Anxiety, melancholy, zest for life, boredom and happiness are examples of basic moods that are not directed at specific objects, but that colour our whole experience of existence. We can observe this upon waking in the morning, before we have even succeeded in directing our attention at any one thing. Moods open the world for us, as Heidegger says, and as long as we are conscious, our relation to the world is disposed in some fashion or other. In this context, we cannot actually distinguish between a person's consciousness and the world by which they are surrounded. Moods are situated in our social and natural environments, not in the individual's heart or mind. Other people cannot feel my headache, but if I exclaim about 'a landscape of mistrust', there is an aspect of that landscape they also can experience. The same can be true of moods within a social or historical zeitgeist.

Because we do not choose them, moods form an existential basic condition of our being human in the world. Nothing demonstrates this fact so clearly as birth. 'She cried when she was born.' Before a being is capable of choosing anything at all, it has a basal 'world experience'. In Heidegger's dramatic terminology, we are thrown into the world, into a place, that is, where we are doomed hereafter to 'throw ourselves out of' the choice possibilities life offers. With that, we can take a closer look at time's ecstatic structure, or at the significance that the future and the past hold for human life.

The past is the mood-ballast with which we are cast into the world. It has to do with our physical attributes, our parents, our surroundings, our society, culture and traditions. These are things that in a fundamental sense make us who we are and about which we can do nothing. And yet the past is not simply a collection of dead facts. If something is my past, it is so because it concerns me in the life I am living now, and the mood from past generations intersects, in a way, with the mood from my further, future life plan. As such we are doubly cast into existence. From birth we find ourselves abruptly in a world of people, and we are doomed to cast ourselves repeatedly, among other people, into new life-projects. We do not choose our past, but the past frames the choices of who we will be at any one point. Indeed, through our relationship to the past we choose who we are.

For Heidegger, the future is often uncomfortably present in the here and now, something that returns us to the ecstatic structure's unchanging character: Godot never comes, which means that the future horizon is never realized as the future. In the future, the present, to be sure, is replaced by a new present. But what makes the future the future is its relationship to the now: that it is not yet realized. As such, the future as future is present in human life.

The past, along with its moods, does not determine what happens to us, but it discloses a space of action possibilities that it is left to us to unleash. We are 'doomed to freedom', as Sartre says. But it is also possible to distinguish between good and bad choices. Good choices are made by people with an eye for that which 'lies in time', who glimpse in a given situation the potential for something genuinely new. In particular, this idea applies to great artists and politicians, people who are able to discern what the present day is ripe for, whether that be a new music form, a new form of law or a new lifestyle. Still, this idea also applies to us ordinary mortals, when we, to the best of our ability, attempt to live up to the place where we have ended in society.

A central component of this societal position is defined by our profession, and a profession brings with it a series of instructions for goal-oriented enterprise. If you are a lawyer, you must work to familiarize yourself with clients, study case documents and practise your speaking skills. It is a matter of goals you set yourself at one point, which are then realized more or less successfully at another time point. First, I acquire my training, and then, at a later stage, comes the baptism by fire in the courtroom. All these activities can be placed in a sequence of present events. What cannot be placed in an equivalent sequence, in contrast, is what my future choices say about my present life: what will my further career be, which clients will I pursue, how will I prioritize between job and family? Such choices concern not only conditions I will realize in the future but the kind of person I will be within the time period before they are realized.

As such, the future not only exists whenever, sometime in the future, it becomes a new present. It also exists in the present by how we choose to relate to it. Because we cannot avoid such choices, our awareness of the future constitutes a weighty horizon. But to be a lawyer and never to make an active, conscious choice is also a (poor) choice about how to be a lawyer. The fact that our relationship to the future determines who we are in the present is something we simply cannot escape.

And somewhere in that future horizon lurks death, not just for lawyers and other professionals, but as the most certain of all universal human facts. To live life now in light of the future means to live it in light of death, which thereby becomes a part of life. As such, mankind's future consciousness is an awareness of our own mortality, signified by the mood of angst. It is not primarily the awareness that we, who are alive now, will not be alive at some future present point. It is rather an awareness of the margins of the very human: for the small light that we, with our ecstatic time, create within the cosmos's eternal dark. Not least this idea applies to history. For, according to Heidegger, we do not have a human past and future consciousness because we have a history. The reverse is the case: we exist as historical beings because we have an ecstatic time structure. History's finitude is identical to the limits of this structure.[9]

THE SPRINGBOARD FOR this chapter was 'the problem of movement'. How can the past and the future be present in the now? Bergson responded that in our consciousness the past permeates the present; Husserl that present consciousness extends to past and future; and Heidegger that past and future consciousness determines who we are in the present. All three discuss the time problematic in relation to man's time experience, and so the issue is whether all time can be explained in this way. For Heidegger's part, the question becomes to what extent our ecstatic time structure applies only to a special 'authentic' human time experience, or whether it can explain the fact that time as such exists.[10] In order to approach this idea, we must first take a closer look at mankind's special form of historical time.

FOUR

RECURRENCE TIME IN LIFE, RELIGION, HISTORY AND LITERATURE

One of summer's joys in Norway is to hear Lars Lillo-Stenberg's 'Neste sommer' (Next Summer) sound from the radio almost every single day:

> If you come again next summer
> I'll be here just the same
> And we'll sing the old songs again
> If you come again next summer
> We'll drink wine just the same
> And talk together about the same old things.

For some of us it is almost a ritual: there is no true summer without this song, which illustrates its own point. The song is not something one grows tired of hearing; to the contrary, the yearly repetition adds an extra quality.

So it is with many childhood delights. Think of the little boy who says to his father, 'Dad, won't you tell that joke?' Then the father tells the joke and both shake with laughter, even though they have heard it hundreds of times before. There is something about the recurrence that creates a joyful pleasure. Is it the certainty of sharing an experience only they have in common? The habitual aspect of recurrence has, at any rate, an inclusive side. It shapes and solidifies a 'we' recognition among those who participate in a practice, something that becomes apparent during major annual, religious and national celebrations. Yet perhaps the temporal aspect is just as important here as the social. Perhaps the repetitions inherent in customs, jokes and other genres are also a visible expression of mankind's deepest relationship to time?

Or take the oral, heroic poetry of previous eras. When the Homeric epics were told and retold, year after year, to the same people, we can assume it instilled a sense of security relative to existence's mutability, something along the lines of, 'In our stories the world does not change, and when we repeat them, the patterns of life stay as constant as the movement of stars across the sky, as the shifting seasons and the biological rhythms of flora and fauna. Telling stories is, furthermore, our appointed task. While we tell them, we are time's masters.' If we turn from literature to the philosophy of life, religion and mythology, it becomes even clearer that the recurrence of time sequences does indeed introduce a particular way to master time.

Godless and God-fearing Repetitions

The father of psychoanalysis, Sigmund Freud (1856–1939), believed he had found a mental repetition compulsion among neurotics. The neurotic repeats instead of remembering, as he so strikingly formulates it.[1] The idea is that painful experiences are first displaced, so they are not recalled, and then the experience is repeated so the individual masters the situation in which he or she was originally a victim. The child abandoned by her mother allows her doll to disappear, only to let it reappear, and beneath the game lies an unconscious desire to overcome the feeling of abandonment. The feeling cannot truly be overcome, however, so the child instead stages the disappearing game again and again. At least in the game the child controls the situation. Freud discovered a similar mechanism in adult obsessive-compulsive neurosis, where these become futile attempts to control a past that will always elude dominance.

It can seem that mankind's general relationship to time is characterized by a similar recurrence pattern. It is a pattern, in any case, that cannot be explained psychologically through specific, negative experiences. The state of being in time, after all, is no definite experience, but instead comprises the basic condition for all experiences. But this basic condition is itself experienced as an existential problem – indeed, as perhaps the most profound of all existential concerns. Therefore, it is not surprising that, even as Freud considered the repetition compulsion to be a response to specific problem situations in time, so repetition has proven a philosophical and religious attempt to tackle or even overcome time. We find the most heroic, non-religious form of this attempt in Friedrich Nietzsche's (1844–1900) notion of the eternal recurrence of all things.[2]

Many people think that no joy endures, and so from an eternal perspective it plays no role in how we live. To the contrary, others say, it plays a huge role in relation to our possible perdition or salvation after death. To both of these perspectives, Nietzsche responds: for you there is no other life than the one you are living on Earth, it is all you have. But imagine living it again and again for all eternity. How would you conduct your life then? You could not put your faith, as do many religious people, in some transfiguration in the afterlife. But you also could not, like the traditional atheist, claim the choice was not all that important. In both cases, a life lived in boredom or frustration will be everlastingly cheerless. Therefore, the mantra of eternal recurrence becomes an appeal to always live your life as you *want* to live it deep down. And, if you cannot live as you will, you can will what you must: say *yes* to your fate!

The idea of positively accepting recurrence as fate is also held by the Danish Christian philosopher Søren Kierkegaard (1813–1855). In work and family life, we are condemned to a daily grind filled with routines and seemingly pointless repetition: day in and day out going to the same job; day in and day out returning home to the same spouse and the same children with the same problems. Yet, Kierkegaard says, throw yourself into this routine body and soul, actively identify yourself with repetitions. For that is your life here and now, and you have no life but the one here and now. Yesterday's first love, of course, can never be repeated as it was, and the hope of 'something entirely new' tomorrow will not be fulfilled if you only dream about it.[3] With a view towards the theme of Chapter Five, we can say that both Kierkegaard and Nietzsche advocated for living in the here and now, not in the past or in the future.

Within Nietzsche's ideas surrounding an eternally recurring time sequence, however, time is, paradoxically enough, overcome. In this case, return does not mean the repetition of things that might potentially change underway. Instead, it is a matter of the same thing returning: 'die ewige Wiederkehr des Gleichen', as Nietzsche and other Germans put it. If we consider what that phrase means, we are again left with an existential life affirmation, while the idea of literal recurrence evaporates. If the idea expresses truth, after all, we would not remember anything from our previous life anyway. Nor could we plan for our next life. Both actions would deviate from the return of the same. Therefore, every single new version of life here and now would lack a past and future consciousness of other instances of the same life. In practice, eternal return is situated here and now – right now. The point is not that everything actually repeats, but that the now acquires an interminable, timeless weight. Ultimately, this staggering thought represents a secular version of traditional, religious conceptions regarding the relationship between time and eternity.

I mentioned hearing the same song every summer as being a near-ritual experience. But there is nonetheless a difference between that and the rituals we find among older and more recent religions. Some rituals have a magical aspect, others many people today would say are based on superstition. What characterizes otherwise differing religious rituals is, first, the repetition of certain actions among believers, and, second, the distinction between ordinary and religious time. Within a shared practice, that is, the succession of events in nature and human life is regularly overcome, allowing one to achieve a direct, timeless contact with the divine powers in existence. This establishes the distinction between sacred and profane time, something of which the Romanian religious historian Mircea Eliade (1907–1986) gives a number of examples from archaic societies in America, Polynesia, Australia and Asia.

Within all such societies, Eliade finds ritual and magical actions to be tied to an eternal return of regenerated life, and celebrations of a new season or new year are often preceded with a chaos phase of wild orgies before the marking of a new order. In the ancient Persian new year celebration of Nowruz, this process occurs through a sacred renewal of the destructive time that has 'fragmented' human life. First, the world's destruction is typified through unbridled feasting. Next, the individual renews himself through participation in the world's creation. This celebration, moreover, is not undertaken in order to reproduce the cosmic creative act but as a manifest reactualization of it, outside of time's normal flow.[4]

Even as ritual festivals unfold as a succession of profane moments, they also occur in eternal, non-successive, sacred time. Participants, as such, consider themselves to literally be one with the god they worship. They can achieve the same thing, further-more, through ordinary actions such as planting, boatmaking, healing, therapy, hunting or fishing. These actions must simply be carried out in specific ways that mark the difference between the sacred and the profane order. When fishermen in New Guinea mimic the mystical hero Kivavia before going to sea, that does not signify their worship of him but rather the fact that they regard themselves as identical to him.

This ritual unity with the divine is not something we only find in archaic societies, but is also clearly present in Christianity. Think of the communion rites. Within them, Christ is considered to be concretely, tangibly present among believers, who are then transported simultaneously into a sacred order with no distinc-tion between past, future and present. Some of the same is true during the Easter celebration when Christ's death and resurrec-tion is made present through the annual, religious observance. Just as it does for Nietzsche the atheist, repetition furnishes the moment with a particular quality that shatters the framework of recurrence in time.

On the other hand, Christianity, Judaism and Islam distinguish themselves from other religions and life views by downplaying the element of cyclical repetition within the cosmos. In Hinduism, for example, the cosmic order is repeated even as nature's cycle repeats from year to year. The world is recreated, in eternity without end, and the same is true for us people, at least so long as we are not redeemed into *atman* beyond time. Within ancient Greek cosmology as well, the world forms were recreated in an endless cycle of purposeful movement. In the Middle East's Abrahamic religions, by contrast, the entire physical world is considered to be created from nothing (*creatio ex nihilo*) at a single and unique past time point. In Christianity, furthermore, Jesus' birth and death on the cross provides the world's course with a clear direction towards possible salvation on Judgment Day. Contrary to the circular time concept inherent in other worldviews, this unique orientation makes Christianity's time linear.

Today, we often regard mankind's secular history as having a linear time flow as well. Such was not always the case in Christianity, however. To Augustine, our world history was as closely tied to the natural cycle as it was to the ancient Greeks. Linear history, for its part, applies to the salvation of the human soul, and salvation history unfolds on the same immaterial and, at its most profound, timeless plane as does contact with Jesus' body during the communion sacraments. The linearity in this time conception is more a symbol in time of something occurring outside of time, if we are to believe the German historian Reinhart Koselleck (1923–2006).[5]

According to Koselleck, the Renaissance is not an expression of any true historical awareness in Christian Europe. Antiquity, however, was regarded as an ancient model for a European 'rebirth' (re-naissance). Using Albrecht Altdorfer's painting *Alexanderschlacht* (The Battle of Alexander at Issus), which the artist completed for Duke Wilhelm IV of Bavaria in 1529, Koselleck demonstrates how ancient heroes, even down to the smallest detail, are depicted as being the Renaissance prince's Christian contemporaries. Alexander's Persian enemies, furthermore, are indistinguishable from the Turks whom that same year besieged Vienna.[6] A number of lifestyle differences from the two thousand years that had passed since the beginning of Greek antiquity were naturally acknowledged. And, obviously, people knew that many important events had succeeded each other in time since then. Yet succession here was not regarded as being qualitatively different from a succession of natural occurrences. In other words, no distinction was made between natural and what we typically call historical time.

Koselleck only identifies a true historical consciousness towards the end of the 1700s. In Europe, the collective singular noun 'history' (*die Geschichte*) did not appear until around 1770, he argues, in contrast to the plural noun 'histories', in the sense of stories (*Geschichten*).[7] Only then did one refer to a specific historical time: a secular time with the same linear direction as salvation history's religious time. Crucial to this dawning historical consciousness was the Enlightenment's belief in progress regarding humanity's development. Through this idea of progress, the circular view of history as it resembles eternal return was abandoned, and the future emerged as an open horizon for genuinely new things in existence.

The new future perspective engenders an equally pervasive transformation in past perspective. Just like for Heidegger, the past appears to be a horizon of significance for active, future-oriented people, not a succession of natural events. As Koselleck observes, the past, from being a collection of former presents, becomes our present past. As such, secular history also becomes something that happens once and only once: history is *our* individual past, not simply some general past caught up in nature's pattern of eternal, constant recurrence. This again introduces a qualitative difference between natural and historical time.

History and Hermeneutics

When the Bosnian-Serbian nationalist Gavrilo Princip woke up on the morning of 28 June 1914, loaded his pistol, and then, pale but composed, wandered into the city streets, did he think, 'Finally, I'm going to fire the Sarajevo shot'? Likely, the thought did not occur to him because 'the Sarajevo shot' is an idea that we, more or less correctly, associate with the start of the First World War. That is something *we* can do following the start of the Second World War. But Princip and his contemporaries could not formulate these ideas, even after that war was well underway. They called the war, among other things, the Great War, the Imperialistic War and the War to End All Wars. But without the ability to foresee the fact that another world war would ensue, they had no reason to say 'the First World War'. Likewise, 'the Sarajevo Shot' is an idea that has only been used retrospectively, that is, after the fact.

This illustrates Koselleck's thesis that an open future introduces a variable past. Not only present meaning but past meaning is created as we go – a central theme in newer hermeneutics, which refers to the theory of interpersonal understanding between texts and other meaningful utterances. Princip's grasp of his own action cannot possibly have been equivalent to our understanding of that same action. The repetition of the phenomenon 'world war' already excludes it. The whole conceptual universe tied to something 'only' being the First World War is, after all, drastically different from the conceptual universe tied to 'the Great War'. These sorts of differences are not found in physics and other natural sciences.

A finger curls around a piece of metal. The metal piece shifts. A lead bullet shoots with violent force out of a tube. After a while, this process is repeated, first once, then hundreds and then millions of times, spreading across the world's continents. With that we have described some key physical events in the First World War, and for a physicist or a biochemist the relationship between finger, trigger and projectile is roughly the same from time to time. But events do not only consist of physical movements. They also have human significance within a context of intentions, customs, institutions, norms and values. In describing the beginning of the First World War, we thus use words like 'heir to the throne', 'assassination', 'power', 'rivalry' and 'imperialism', not just 'finger', 'adrenaline level' and 'mass deaths'.

Is it not the case, however, that parts of inanimate nature also first acquire meaning within a larger whole? Perhaps we are ambling the smooth rocky slopes by the sea and suddenly we spy a couple of water drops behind an outcrop. For a split second we do not know if it is just a random bit of foam spray or the beginning of a wave before the rest of the surge rumbles against the rocks. Two or three isolated drops do not immediately constitute a wave, and only after a certain amount of time can it be established that, yes, they did just that. The whole into which the water drops are incorporated determines the kind they were from the start. Still, this has nothing to do with the type of hermeneutics we are talking about here, which is limited to human actions.

We describe what we see with concepts like 'water molecules', 'drops' or 'waves', but the molecules, drops and waves do not describe themselves. Unlike us, they have no ideas about what is happening, and they have no self-comprehension that can depart from our understanding of them. What is distinctive about human actions is that not only do they enter into a larger, meaningful context that ties things together and can be retroactively described from without. Long before history took its course, people in former times retained a future horizon that encompassed our present, just like our past horizon encompasses them. Without awareness, lifeless things cannot reach out to each other in this way. But we do just that with our thoughts. As such, we engage in a mental dialogue with other people and take a position on what they think and do.

We cannot take a position, however, on what happens with a water droplet. We cannot assess its movements as noble, low, adorable, inspired, careless or logical. But if we perceive other people's actions in light of our own future projects, we cannot evade an element of judgement. That is, hermeneutics is not a theoretical, scientific method, but a practical reflection on how one's own rational, emotional, moral and aesthetic development might provide new insight into past actions. If we begin, for example, to consider secular principles of law to be civilizational progress, then this viewpoint can yield a deeper understanding of life in earlier, autocratic or fundamentalist cultures. In other cases, of course, our own time can represent civilizational regress. In both cases, this historical understanding is normative and appraising, not just value neutral and descriptive.

And with that we have approached the previous century's most prominent hermeneutical thinker, Hans-Georg Gadamer (1900–2002). Koselleck was a student of Gadamer, and Gadamer was a student of Heidegger. From Heidegger he learned that all our understanding of the past is characterized by our life's outlook for the future. Severed from such practical engagement, history loses its human meaning. That means that our overall relationship to history does not essentially differ from our relationship to works of art, according to Gadamer.

Musical, literary and pictorial artworks are (1) created by specific people under specific conditions, then (2) are placed in the world as available for interpretation and finally (3) are interpreted by the public or by practising artists. This tripartite structure introduces an issue recognized by all who have had dealt with music and other art forms: should the practitioner remain as 'true' to the work as possible, and does that mean immersing oneself in the composer's state of mind during the creative moment?

This last point seems irrelevant. Perhaps Mozart was in a foul mood while working on a symphony after a quarrel with his Constanze, and maybe the new composition was a way to overcome his distress. In that case, we would still not interpret the musical intention in light of his mood but in light of the work. Should we not, therefore, adhere to the work as we see it before our eyes on the page, interpreted in light of past conventions for musical notation?

Even then the result would not be a recreation of the 'original work'. Instruments have developed further and the same is true for our culturally conditioned ear. Therefore, a tangential movement that back in its day sounded youthful and fresh can today sound antiquated and stale. Furthermore, repetition itself changes what it recreates, and conscious repetition can even transform a spectacular first-time experience to kitsch. If we are to convey Mozart's musical universe, we must consequently present the work in a way that differs physically from what the composer himself did. But how? In a way that gives us a musical experience today.

According to Gadamer, this idea signifies that our relationship to artwork always has a creative side. Taking our own musical sense as a springboard, we must present the work as it appeals to us. That means that if we understand the original work, we simultaneously judge its calibre. Without such evaluation, we would be unable to discern what qualities it is possible to retrieve from it. Uncovering these qualities undoubtedly demands respect for the composer and for the tradition that has shaped our musical sense, so that we approach the work with a certain pre-understanding. But even if the composer be ever so great, his or her original experience cannot directly be recreated under subsequent, altered conditions. It is rather that the 'work' encompasses both the composer's and later practitioners' musical engagement. In Mozart's case, perhaps we ought to strive to play in a manner we presume he would have liked, had he lived today?

It is this approach to artwork that Gadamer believes must necessarily characterize every understanding of previous events. As with the composer's 'original work', so the meaning of past actions is not something that simply resides there, ready to be excavated. Even as sonatas incorporate both composition and subsequent piano performances, so the meaning of historical events incorporates both the actor's intentions and the historian's interpretation in light of his or her present engagement.

In a way, every historical epoch contains the significance of all of history. On the one hand, this means that the full and complete meaning of an individual action can only emerge at history's end, that is, never, so long as there are historians. On the other hand, an action's significance becomes, in retrospect, steadily richer. The more historical experiences to which it stands in relation, the greater becomes its present, meaning-giving context. If, furthermore, historical significance is not as subjective as the individual's fear of death, for example, then it is also something about which it is possible to have a more or less adequate, rationally based knowledge. Heidegger's ecstatic time structure acquires thereby a social and not merely an individual–existential validity.

As we have seen, this time structure means that time's three modes, past, future and present, do not follow each other as links in a chain of events.[8] From the perspective of the now, and because the future exists only as the future and the past as the past, both modes are in a way timelessly present at once. In light of history's hermeneutics, we can explain it like this: as physical events, human body movements succeed each other in the same way as do other physical occurrences. Historians will always have as one of their tasks to study such event progressions. But if we look at the significance of those same actions, then the future has retroactive force. It changes the past, almost as if we moved every which way across time. Is something like this even possible? It happens in the narrative time of stories, at any rate.

History's Stories

The Sarajevo shot is not just a body movement that enters into a network of physical causes and effects. It also has value-laden human meaning. Is this not also the case with other human actions, not just with unique feats like unleashing world wars? As we have already discussed, we always perceive what we do in light of social norms, institutions and traditions. My wish to take an exam, find a secure job or marry before a certain date would be impossible without exam regulations, a legally structured work life and the institution of marriage. Without such social environments, we have only action fragments in the form of episodic, single occurrences.

The English moral philosopher Alasdair MacIntyre (b. 1929) illustrates this difference between actions and action fragments by citing the following travel report:

> There we waited on the ladies – Morville's. – Spain. Country towns all beggars. At Dijon he could not find the way to Orleans. – Cross roads of France very bad – Five soldiers – Women – Soldiers escaped – The Colonel would not lose five men for the sake of one woman – The magistrate cannot seize a soldier but by the Colonel's permission, etc., etc.[9]

We suspect there is an account of human action here, but it is just a suspicion. In order to grasp the meaning behind the snippets, we must understand more than just a snippet. According to MacIntyre, we must flesh out the report so that it forms a true story: a story regarding the context of a behaviour that only then assumes the character of being an 'action'. He favours, therefore, a narrative action theory: that which distinguishes specific human actions from reflexes or other more or less mechanical behaviour is the stories into which such body movements enter.

In order to grasp the appeal of this kind of action theory, we might recall the various narrative elements that from childhood on became part of our socialization. We listened to fairy tales by the bedside, we watched children's television shows, we read kids' books and we immersed ourselves in the lives of heroes and villains at the cinema. Other personality-shaping stories deal with prominent women and men in our national past. Our self-understanding is thus formed in light of actual and fictional narrative patterns, a formation that continues into adult age and never ceases to place its stamp on our actions. For, as self-conscious beings, we humans never act merely to realize specific desires. We are also perpetually aware of what kind of people we are when we act to satisfy these desires; if, for example, we happen to cheat on our taxes to pay a bill.

MacIntyre argues that it is this ability to act in light of a narratively formed self-understanding that makes us human: man is essentially a story-telling animal.[10] By that, he means not only that we find ourselves, on the one hand, in real history, and, on the other hand, tell stories that are more or less fictional. Stories can harbour deep truths because history in turn has a narrative structure determined by our tales. This does not serve to erase the distinction between fiction and fact. Yet factual history is the history of beings who also perpetually live their lives in light of tales with more or less fictional elements.

If MacIntyre is correct here, it can have something essential to say about time and our relationship to it. Lacking in the above action fragments is a structure that binds everything together as a unity. What we have also previously seen missing in a sequential time composed of isolated events is a structure that binds them together in a unity – of past, future and present. So, if the stories in our lives create unity in our actions, perhaps they also create unity in time. In his three-volume work *Time and Narrative,* the Frenchman Paul Ricoeur (1913–2005) argues that this is, in fact, the case.[11] In order to uncover unity in time's phases, he abandons purely philosophical time theories and seeks instead a 'poetic answer' in Aristotle's *Poetics*.

Central to Aristotle's literary theory is the concept of mimesis, that is, the imitation of human life. Contrary to a passive mirror reflection of the world, the poet's 'mimetic' act is a vital reproduction of human activity. The subject of imitation is people engaged in actions, and both the poet's creative deed and the audience's later appropriation of the work are considered actions as well. So perhaps 'imitative representation' is the most adequate translation of the Aristotelian concept of mimesis. For the idea centres on the artistic depiction of people who themselves represent things in word and deed, or on the narrative reproduction of things that happened in the storyteller's life, as we just saw with MacIntyre.

For Ricoeur, the unity of representing and represented life means that tragedies and other stories have a tripartite structure with clear parallels in Gadamer's hermeneutics. In the first place, we have mankind's tradition-imparted, narrative understanding of ourselves and our actions. This means, Ricoeur argues, that the poetic work is prefigured by conditions external to the work itself. In the second place, the work also creates a configured order of the prefigured elements in life that lack a manifest order. In the third place, human actions are refigured in the reader's and audience's subsequent life. The time philosophical implications of these three ideas is the following: our daily, prefigured life contains narratives, often with poetic elements, even. But as we all know, life can be chaotic, complex and full of unforeseen things that break in and frustrate meaning. Sometimes these disruptive elements bring incomprehensible tragedies, as when we are struck by illness, accidents and death. For even if all actions play out against a cultural backdrop of linguistic, symbolic and institutional meaning, life is full of things that 'just happen'. The cultural backdrop forms a general framework that also encompasses isolated events that lack any obvious significance. All such isolated events, furthermore, occur in a time sequence of before-and-after relationships. However, without a narrative, ordering whole, this sequence is as episodic as a logbook-like travel account. All of this is transformed, however, through the poet's eye for general, meaning-giving structures in daily life.

Yet why do we, as the audience, enjoy a good drama on the refigurative level? According to Aristotle and Ricoeur, it has to do with the joy of understanding something, of having a new, meaningful insight. The pleasure becomes even greater if our apprehension is a form of self-understanding. Through fear and sympathy, we identify with the tragic hero's life and way of living. Without being devastated by the drama ourselves, the sudden insight into what created a person's tragedy can exercise a liberating effect upon us; when we, for example, see a person driven into the abyss as a result of an apparently insignificant character flaw. Then we recognize a common, configurative pattern in our own lives, and this narrative ability to glimpse and represent such patterns is what Ricoeur calls the capacity to form plot.

Through plot, the threads within life's drama are collected; through plot, they gain a direction and significance; and through plot, they thereby overcome the action sequence's episodic character. As audience to a new drama, we often grasp the plot only towards the end, and then with an even more startling insight that – given the social environment, the external conditions and the hero's character traits – it 'must end like that'. At that point, we see that events we thought were near-randomly piled atop each other were not structured that way after all. They pointed forwards, towards a momentous and unforgiving but comprehensible and, in that sense, meaningful end. In this way, plot's narrative form creates a retrospective unity in otherwise disparate, episodic chains of events, something that, first of all, has an obvious aesthetic function. Second of all, it becomes Ricoeur's 'poetic solution' to the time philosophical issue of binding together past, present and future.

Not only does the viewer or reader experience a retrospective, memory-based insight into how the plot must end. In narrative time, we also move freely between fiction and reality, and on the fictive plane we can shorten, expand, omit, turn and reshuffle the elements in a recounted event sequence. We can take significant time in representing a brief event sequence and also the reverse. Not least in narrative we can return to the beginning from the end, in a time that seems to flow backwards. As such, we repeat something that has already been recognized but not understood. The result is both a literary experience of time's episodic character and 'an alternative to the representation of time as flowing from the past towards the future'.[12]

This type of recurrence within a story attains its full effect first through another kind of recurrence: when the same story is repeated. In this case, the beginning is retrospectively permeated by a meaning the audience already recognizes:

> I just spoke of the 'end point' as the point from where the story can be seen as a whole. I may now add that it is in the act of retelling rather than in that of telling that this structural function of closure can be discerned. As

soon as a story is well known – and this is the case for most traditional or popular narratives, as well as for those national chronicles reporting the founding events of a given community – to follow the story is not so much to enclose its surprises and discoveries within our recognition of the meaning attached to the story, as to apprehend the episodes which are themselves well known as leading to this end. A new quality of time emerges from this understanding.[13]

With that, *we* are also returned to where we started in this chapter. Literary repetition transforms time into something we govern, almost into a joyful play, as people in Homer's day experienced the oral performance of *The Odyssey* and *The Iliad*. Ricoeur's narrative, 'poetic solution' to the problem of the relationship of natural scientific and human time to one another shows that it is possible to regard narrative as an activity anchored in time's most essential character. It is doubtful, however, that a theoretically oriented reader will be content with this attempted solution, and in the next chapter we will acquaint ourselves with a more systematic discussion of the time concept within the natural sciences.

FIVE

IS THERE ANY TIME OUT THERE?

In 1908 an article called 'The Unreality of Time' was published by an English philosopher with the eccentric name John McTaggart Ellis McTaggart (1866–1925). His habits were no less eccentric, since apparently he tended to invite friends and colleagues to breakfast without offering them actual food. Perhaps he considered food for thought to be adequate fare? That would have suited his basic philosophical viewpoint, which he adopted from Georg Friedrich Hegel (1770–1831) and other idealists who argue that, at its most profound, existence is immaterial and rational, identical with logic in human thought.[1] Meanwhile, it was within analytical, scientifically oriented philosophy that McTaggart especially made his mark, arguing a thesis even more eccentric than his name and breakfast habits: time does not exist, it is an illusion.

By claiming that time is 'nothing', McTaggart promoted what we can call a form of time nihilism (Latin: *nihil* = 'nothing'). The Greek Parmenides (*c.* 515–450 BC) did the same thing, and McTaggart continues to be Parmenides' greatest advocate in modern times. The philosophers we encountered in Chapter Three, on the other hand, do not deny time's reality, though they do tend to say that time does not exist in the world beyond mankind or other conscious beings. Perhaps the most important precursors to this position, which is often called time idealism, are Plotinus (AD 204–270), Augustine and Immanuel Kant (1724–1804).

McTaggart's 1908 article has put grey hair on the heads of those thinkers who wish to preserve a reasonably good relationship to time, and in this chapter I will try to show the reader why the arguments McTaggart makes are actually quite convincing. As a stimulus for the purely theoretical considerations, however, I will take up first the thread from this book's introductory pages, with some further reflections on time's intangible character. Perhaps with that it will be easier to understand how someone can come to the apparently insane conclusion that time does not exist. After a discussion of McTaggart's time nihilism, I will then sketch a possible time idealistic alternative, which will lead us to the final chapter.

Time's Disappearing Act

It is difficult to pin down time, either as something concrete or as something abstract; whether it is something we sense or something we imagine. We can see, hear, smell, taste and touch things in the world, but we are not equipped with any one sensory organ for time. As Thomas Mann's (1875–1955) protagonist Hans Castorp says in the great 'novel of time', *The Magic Mountain*:

> 'Just be quiet. My mind is very clear today. So then, what is time?' Hans Castorp asked, bending the tip of his nose so forcefully to one side that it turned white and bloodless.

'Will you please tell me that? We perceive space with our senses, with vision and touch. But what is the organ for our sense of time? Would you please tell me that? You see, you're stuck. But how are we ever going to measure something about which, precisely speaking, we know nothing at all . . .'[2]

What is said here about the difference between time and space is perhaps not so obvious, but one thing seems clear: time has no particular sensory quality. And yet it can permeate every sense – whether it be with smells, colours, touches, tastes or sounds – by an intangible form of extension. Objects with form, shape and size have a spatial extension, and if they endure for a certain time period, it is almost as if time is trying to imitate space with its own particular brand of extension. But it just cannot quite pull it off. For whereas space stretches like a languid cat before our eyes, time merely stretches into a disappearing past.

Here we can imagine a short dialogue. 'You just don't give up, do you?' space says to time. 'You'll never be able to stretch, but you'll never stop trying. What, do you just want to be like me?' 'Maybe,' time responds, 'but I also have higher goals than you. I want to reconcile the irreconcilable and make people both young and old – at different times. And I can do magic. Just like that, I make the now appear, and just like that, I make it disappear.'

Space is right, in any case, that time has issues with its own extended duration. Since the future has not yet arrived and the past perpetually vanishes, we must enlist the aid of space to help conceptualize it. As Kant in his own day pointed out, if we draw a line in space, we can visualize that line as a temporal sequence. But we cannot envision a 'time-extension' by itself, and as an image of time's passage, it is as if the timeline is borrowing from space, with past and future placed outside each other.[3]

But if extended time and space segments are not sensory qualities, they are not abstract thought-objects either. There is no physical limit to what we can imagine, after all, and neither time nor space hinders a logician's or mathematician's intellectual fantasies and ability to abstract. Numbers, too, have no actual extension in time and space. Of course, there is often talk of pure mathematical 'space', for example, with probability curves. But that does not mean the space in which concrete objects have extension. In the same way, we can imagine mathematical 'times', thereby disregarding the specific duration we typically associate with a time sequence.

Something similar occurs within stories, in so-called narrative time. If we say, 'And then three years passed without anything significant happening,' we can overlook the actual duration of a year. Or take a science-fiction story from planet X where we know that ten time units of 'mex' equals one 'tex'. If we know that a hundred tex have passed, we are talking about mathematical time perception. Nonetheless, we have no idea how much time has passed if we do not also know the duration of one unit. We know that the unit has this or that duration, and we can imagine that such and such length of time has passed. But the imagined and actual duration of things are not the same, and if we want to know how much time something takes, we must connect it to sensory details. Yet we have seen that time is itself not a sensory quality. No wonder it is so difficult to grasp!

Furthermore, time is more intangible even than space (which is also located at a strange intermediate position between sensing and thinking). For change happens in time, and time's moments cannot be confined; they slip away. Therefore, time is typically symbolized with something that flows or with fire or some other transitory phenomena. So fleeting is time that one can call its very existence into doubt. Yet, the phenomenon of change also makes it difficult to doubt time's existence. How can change exist without time?

Perhaps we are watching as a ball rolls from point A to point B. If we had only three-dimensional space and no time, we would have to say that the ball was at both A and B, and that seems like a contradiction. But if we argue that an object cannot be in two places, we imply a short addendum: 'at the same time'. An object cannot be in two places simultaneously, only at two differing time points. So it is reasonable to believe that time makes change possible, and hardly anything has a more conspicuous reality than the changes existence brings: the leaf carried away by the wind, the silence broken by a scream or the child grown into adulthood. A leaf cannot be in multiple places, a person cannot both hear and not hear a scream and a person cannot be both adult and child. We assume, however, that all these things are possible at different moments in time. As distant from 'nothing' as it can get, time must instead be one of the most actual things out there. Yet it is fleeting; as fleeting as the changes it helps keep alive.

This fleeting aspect makes time, as Thomas Mann indicated, problematic to measure. In space, by contrast, measurement is a simple and straightforward affair. We have a plank, we have a metrestick and we measure the plank's length by placing the metrestick along it a certain number of times. We know the metrestick's length because we can hold it between our hands. Indeed, many units of measurement originally referenced human body parts or body movements – 'foot', 'pace', 'fathom' (which comes from Old English *fæthm* or 'the arms outstretched'). Even if we were alone in the cosmos, we would still have a perception of size using our body as a measuring stick or unit of space. But recognizing the measuring stick in this way yields no real knowledge of objective magnitude. That we discover only by observing our body and its movements relative to other world objects, so we learn how exactly we can sit or lie, stand or walk on them. What about time measurements?

We observe the relationship between the metre stick and the plank when they touch each other, something we see as clearly as either metrestick or board alone. In this way, we are using one part of space to measure another part of space, and this type of side-by-side or edge-to-edge measurement has no parallel within time. We cannot use one part of time to measure another part of time, which, in the meantime, has disappeared. That would be like waving our arm once, then again, in order to see how long the first wave lasted. The first wave is lost for all time and can be made present only through memory.

This problem is grounded in the fact that we cannot place one time segment against another, as we spatially do with length measurements. If we set the short ends of plank and metrestick against each other, we would not have measured any length. But if we want to measure two time segments against each other, all we have in a way are short ends. For, left to itself, time has no room for multiple parts placed against each other, but only following one after the other. Without space, time is narrow as a streak, not even as wide as a plank. As such, two time segments can only touch at a single point, without any measurable extension.

The case is different when observing the rings in tree trunks. They are attractively arranged next to each other in space. And in space we see that some are thin and some are thick. Furthermore, we know that time has added a year for each ring. If we count the rings, we are thereby counting a sum of years. But counting and measuring are not the same thing, and something we will never achieve is having the units in one time sequence measure the units in another time sequence.

But what about clocks? Do we not use them to measure time? As we saw in Chapter One, there is nothing we modern folk measure so often as time, so does this fact not contradict what I just said? No, because even if a clock hand, contrary to a tree ring, moves, we do not use it to measure something that will happen or has happened, we use it to measure something happening right then. A digital clock's number sequence does not show the duration of past number sequences, but something that is occurring in another place. As we have seen, clocks serve to synchronize our activities with things happening at different points in space, not to measure things happening at other points in time.

This lack of concrete time measurement led the German philosopher Gottfried Wilhelm von Leibniz (1646–1716) to wish he could 'keep a past day for comparison with days to come, as we keep measures of space'.[4] Augustine, however, provides the classic analysis of the principle limitations on time measurement. If time exists, it can be measured, he thought, and because the duration of a past or future time unit cannot be measured, the result seems to be a form of time nihilism.

Whatever else one thinks about time, it consists, according to Augustine, of extended units with three phases: past, future and present. But the future does not yet exist. And the past no longer exists. Therefore, the only reality is the present. But if we look closer, the present also dissolves. No matter how small we make a time unit, still it has duration; it consists of the three phases. In this way, the present shrinks into a tiny disappearing element between a non-existent past and an equally non-existent future. It becomes an imaginary frontier between nothing and nothing. If it prevails, it is without length, and if it has length, it does not prevail. Time in extension vanishes.

Nonetheless, that is not Augustine's last word on the matter; he does not end by denying time's reality. The present is real, and the present can in fact be expanded to include a past and future dimension. But it happens only in our consciousness. The future exists – in our present expectations. The past exists – in our present remembrance. Everything is located in our present consciousness, something that transforms Augustine into a time idealist and into a prominent, time philosophical presentist: only the now, the present, is real.

Still, if everything is only located in our present consciousness, what about changes in existence? If we see a bus driving by, we perceive it first at one place and then at another. Both its first and second positions enter into that movement, and if the process decomposes into a series of nows, we no longer have any movement.[5] If this idea is true, presentism also seems to result in the phenomenon of movement's disintegration. How can time be other than a series of nows if neither the past nor the future exist? It is such questions that McTaggart tackles, both generally and in principle. That is, he does not just take up movement as we perceive it, but all conceivable change in general.

McTaggart on Time as Illusion

We have seen that without time change becomes self-contradictory: an object cannot be in two different locations, both high and low at once, unless we add the fact that this does not happen simultaneously. That time first enables change is something most thinkers accept as a given. McTaggart also adds to this the reverse: without change, no time. There would be no time in a world where all change was 'frozen'. In his 1908 article, 'The Unreality of Time', he argues that change is something that cannot be rationally comprehended, certainly not in relation to time. In this way, he means to expose time and change as pure illusions, at least as these phenomena are described within the natural sciences.

McTaggart takes as his springboard the way in which physics places events relative to a mathematical ordering of time points along a time axis. As we saw in Chapter Two on Einstein, this placement means that all physical events can be situated before, after or simultaneous with each other, such as Holy Saturday 2010 comes after New Year's Eve, before Pentecost and simultaneous with Per's birthday. This succession of before and after relationships is what McTaggart calls the B-series. Also in the same chapter, we saw that, outside of physics, events can further be determined as past, present or future. Right now it is New Year's Eve; in the past, four months ago, it was summer; and in the future, in April, it will be Easter. Ordered this way, events enter what McTaggart calls the A-series. In time philosophical discussions, he argues, we must always distinguish clearly between these two series, and he does that himself in his argumentation that time is an illusion.

The B-series time axis is as immutable as the number series 1, 2, 3, 4, 5, 6, 7 . . . And the number series is immutable because the numbers are and remain separate entities that can never be transformed into each other. The number 4, for example, can never become the number 5. The same is true of the B-series mathematical ordering of time points. The third time point in the row can never be transformed into the fourth, they are and remain distinct.[6] Thus the B-series is unchanging, and so cannot, according to McTaggart, be used to explain how change exists. Only if something *happens* in relation to an immutable series do we have actual change. This is the case if a person is born on a certain date, or if the clock hands move in relation to the numbers on the dial. But an immutable relationship to an immutable sequence does not provide any change.

Or take again the series of New Year's Eve, Easter and Pentecost. In the B-series, these occasions can never be transformed into each other, they are as distinct as the time points to which they are placed in relation. This is an aspect they share with different portions of space, which also form distinct parts of the world. Within such a series, there is room at most for a type of quasi-change, like when we say that a road segment shifts from east to west. We can certainly call this a change, without thinking that an actual change in the road, and not just in our relationship to it, has occurred. Similarly, if we imagine that we are moving relative to a number sequence, what changes is our relationship to the sequence, not the sequence itself. If, in contrast, a certain part of the road grows over, an actual change has taken place. For then one and the same part of the world has altered character. This will never happen, though, in a B-series of disparate events.

If time makes change possible, it cannot be through the B-series and its before-and-after relationships, so we are back to the beginning, to the question, namely, of how one and the same part of the world can be assigned incompatible properties (such as being overgrown and not being overgrown). It is here that the A-series comes into play, because we like to think that something can have contradictory aspects within different time modes: first in the future, then in the now and finally in the past. As such, we conclude that a flower can be both fresh and faded. It fades first in the future, then in the present and then in the past. But that fact does not solve any problems, McTaggart argues. Not even the two most obvious ways of considering the A-series results in any change.

(1) We can say that events are future, present or past only relative to each other in the time series. Thus Easter can be termed future in relation to New Year's Eve and past in relation to Pentecost. However, the placement of these holidays in relation to each other never changes, and so all we have is a refiguring of the fact that they are located before or after each other in the B-series. (2) Therefore any change in the A-series must occur in relation to past, future and present as immutable attributes outside the time series. Yet the point of explaining change with time was that objects could have antithetical properties at different time points. Relative to a timeless A-series, McTaggart argues, it is as contradictory to say that something is both past and future as it is to say that an apple is both rotten and ripe – and that is that.

Not convinced? Perhaps that is because you are left with the sense that McTaggart is overlooking something: that past, future and present have different reality statuses. If reality holds everything together, including that which is present, future and past, then perhaps the idea does become self-contradictory. But can the future not exist as a possible state, even if it is not real? This seems as unproblematic as saying that a short man could be tall. He is not tall, but we can imagine that he might be, as a possibility. In this way, we can also imagine that the future and the past might be real, even if they are not. They exist as possible states. If the A-series explains change in this way, it presupposes, therefore, a form of presentism.

But so far nothing indicates that presentism enables us to meet McTaggart's time philosophical challenge. Augustine's presentism itself encountered the problem of change, and far from resolving the issue, a sharp distinction between possibility and reality will only enhance it: if change is a reality in the A-series, then the transition from future to present and past must also have a certain reality, and if neither future nor past exist – that is, if they are completely illusory – then such transition cannot have any reality at all.

If only time explains that change is possible, then the conclusion becomes that change does not exist, it only seems that way. Since time, furthermore, presupposes change, then time itself does not exist, it only seems that way. Is McTaggart correct in saying, therefore, that time is an illusion, a fantasy anchored in human nature, but a fantasy nonetheless? Those of us who do not wish to steer this theoretical course entirely are compelled to say something about the reality of past and future. Within presentism's null-vision, these time phases have no reality and, therefore, there is no change. There is no change, furthermore, even if they do have reality, for then change appears like a self-contradiction. But what about if we distinguish between different types or degrees of reality? Even if only the present is fully real, perhaps the past and the future also have a certain reality, not only at another time point but in the present? And if so, what does that mean?

Well, for one thing, we humans know something about the way past and future can be present to us. We have an individual and a collective history, as well as a future horizon, all of which helps determine what we do in the present. Bergson, Husserl and Heidegger have all contributed to describing the role this horizon plays in human life, and McTaggart's presentation gives their efforts increased relevance. If mankind's past and future have a certain reality in the present, it is reasonable to claim that every physical change happens relative to human time. In this case, we are saved from McTaggart's time nihilism by a form of time idealism, that is, by the idea that time does not exist outside of consciousness. Does that sound strange? Within European philosophy, at any rate, there is a strong element of time idealism, with the neo-Platonic Plotinus acting as a kind of forebear.

An unconscious universe was, for Plotinus, a collection of characteristics with no inner contradictions. Conscious creatures, however, can desire things and desire is a type of lack; a relation to what one does not have. If you are hungry, you are not full of food, but still food has a certain reality. Desires occur in the present, after all, and concern something you want in the future, and so reality is in conflict with itself. Later this thinking was further developed by Leibniz. Like McTaggart, Leibniz was a rationalist, and like McTaggart, he argued that in a contradiction-free world there was no time. Time signifies change which signifies contradiction. Nonetheless, 'deep down' the world contains no contradiction and so no time. Only for desiring, conscious beings do time, change and contradiction prove a reality.[7]

If we thereby connect time to consciousness, to how people, dogs and other animals can experience a world of shifting impressions and wants, does that truly mean that a world with no consciousness is a world with no time? Yes, perhaps we must accept that conclusion. But it must also be nuanced in light of man's particular forms of action, desire and time consciousness.

Man, Dog and Time

I assume that most dog owners have had the same experience as myself: you cannot schedule things with your dog. I will never seriously say to my Molly, 'Let's meet here at sundown.' Molly has no idea what I mean, and it is not owing to her limited ability to produce or comprehend human sounds. It is because she cannot relate to the future as we humans do when we schedule things, make promises or enter into contracts. A dog's awareness of the past is also different. Dogs can be physically and mentally affected by things that have happened, but they do not sit and chat about them, and nothing shows that they pause to consider the logic of a days-old or weeks-old decision. Nothing in their behaviour, furthermore, shows that they maintain the same thoughts over time and in new surroundings.

But we humans do that constantly, for example, when we think, 'Finally, the day I've been anticipating so long has arrived.' I relate to that day as 'today', though I know earlier I related to it as 'soon' or 'in four weeks', and, in order to do that, I must be capable of virtually assuming past positions in time and viewing the day, which is 'today', from there. Dogs do not seem to dwell on a single, identical concept in this way. But they are obviously conscious beings, both when it comes to respite and to change.

Much of what philosophers have written on the present existence of past and future must, therefore, apply to dogs and many other animals as well.

Take Bergson: in Molly's dreams, the cats she has chased and the dry, bland biscuit she ate for dinner make an appearance. As such, the past remains in her consciousness. Or Husserl: when Molly sees a large dog approaching, she is intensely aware of its next step, interpreted in light of its last one. As such, her perceptual impression is determined by a protension and a retention. When it comes to Heidegger, though, Molly encounters significant problems. Her self-understanding is not dictated by career choices and other long-term projects. Much less does she comprehend the past horizon's historic–hermeneutic character. In order to relate to that idea, it is not enough to be aware of change. One must also be a physical creature that both shifts their position in time and retains facts independent of time's passage.

In Chapter Three, we saw that the human experience of acting in a particular situation is not contained in physics time. Even though we know the physical time point at which something happens, we do not know when it happens if we do not also know when that time point becomes 'now'. We can realize that the camp closes at 10 p.m. without knowing that moment is 'now'.[8] As such, the A-series's present is related to our own, personal place in time; to where we find ourselves right now and not, for example, in a few minutes, hours, days or years. For even if all positions in time constitute 'a now', still you indicate the time point right now if you say 'now'. The same is true of words like 'soon', 'in an hour' and 'last year'. It is 'always soon' and 'always in an hour' relative to some point in time. But if you say 'in an hour', you thereby exclude all other hours.

With that we have identified a central hallmark of so-called indexical words and elements of thought, which also includes personal pronouns like 'I', 'you' and 'she' and spatial adverbs like 'here' and 'there'. Even if all locations can be considered 'here', I exclusively mean the place right *here* when I say 'here', and even if all people constitute an 'I', I only mean *me* when I say 'I'. Grasping this idea is more complicated than we often imagine, and children learning to speak typically use their own names, such as Karl and Kari, before they use personal pronouns.

What further complicates the matter is not only that the same words shift reference from situation to situation. Indexical words are components in a system that does the reverse, making it possible to say the same thing in different situations. They just replace each other according to specific rules. If I say, 'I have a pain in my foot,' another person can say the same thing by switching 'I' with 'you': 'You have a pain in your foot.' If I say, 'Here I am,' another person can say the same with, 'There he is.' If someone says, 'It's raining now,' a person can later say the same with, 'It was raining then.' As such, we humans are capable of retaining facts across time and space, and a similar time consciousness does not seem to exist in dogs and other animals.

A dog's experience is always changing, sometimes by leaps and bounds. In specific, perceptible surroundings, Molly relates to a single time point with shifting attitudes, such as when she rushes at a mole behind a tree, only to discover that there was, in fact, no mole there. Momentarily confounded, she experiences that the impression and the retention do not correspond to the protension. But she evidences no similar behaviour apart from her perceptible surroundings. We humans, meanwhile, can predict things far into the future. We can make claims about what is happening far away, and, sooner or later, our predictions and claims are confirmed or disproven. In this way, we continue to relate to the same time point, which merely shifts mode from future to present along the way.

This indexical reference to reality enables us to exist independent of our changing, subjective attitudes towards reality. If I say or think, 'Now it is finally 17 May,' the national day of Norway, I might be wrong. Maybe I slept the whole day away, so it is already 18 May. But even if all my perceptions of now are shaped by 17 May (flags, parades, sausage eating), it is nonetheless not these perceptions that determine to which day I am referring. With 'now' I am exclusively referring to 18 May, simply because that is the day I invoked. And with the phrase 'in a week', I am referring to 25 May, not to 24 May. Owing to the fact that 'now' and other indexical words and expressions signify events and things independent of our altering perceptions of them, philosophers of language say they have a direct (or a 'rigid') reality reference.

Therefore, we can answer an obvious objection to the tradition from Plotinus and Augustine, namely, that all future reality dwells in our present attitudes. The limitations inherent in dogs' and other animals' future consciousness, however, could lead one to believe that something important is lacking in this type of time philosophy. If a future time point exists only in conscious creatures' wants or expectations, then that same time point cannot, for example, later prove the subject of disappointment. Terminating the expectation would thereby take the particular time point with it, something contrary to the simple fact that a single time point can alternatively appear as past, present and future. But this last idea is precisely what occurs within mankind's indexical reality reference, which does not change with changing conditions. If it did, we would be no more capable than dogs of making appointments or promising to do things.

On 16 May I might succeed in preparing Molly for 17 May by reminding her of past ideas regarding a day filled with splendid sausage consumption. Nonetheless, we would be incapable of promising each other to do various things that day, for example, to eat sausages. If Molly loses her taste for sausage, I cannot reproach her the following day with, 'But you promised we would eat sausage together.' If Molly's desires change and she loses her taste for sausage, her relationship to the future changes, and there is nothing more to be said. So long as her entire connection to the future resides in her present desires, she can in no way tie them to any future goal. Nor can I appeal to a promise. All I can do is try to convince her to like sausages again.

Some people will perhaps argue here that human promises are essentially an expression of specific, present wants, and that to bind oneself in the future is simply to emphasize certain, higher-ranking wishes at the expense of others. And I am unable to directly disprove this objection. But one condition for 'true' promises seems to me obviously met by mankind alone: we can relate to the future ('I promise . . .') independent of our present wants and ideas. The same is true of the past, so that, independent of our present experiences and memories, we can hold one another responsible for previous actions ('You promised . . .').

We can also speak *now* about things that happened *then* or will happen *soon*, and so our thoughts can themselves potentially encompass endlessly more than that which via memory and expectation affects our actions. Indeed, both in time and in space we can imagine lines projecting to any part of the universe, and, equipped with maps or spacetime diagrams, we can let a dice roll determine the coordinates of positions to which we have no practical relation. Still, we understand what it means to go to these places. Even if the dice show the coordinates for 'Sydney in three months' and we have no desire to travel there in the foreseeable future, still we can talk about what it would be like to go there. Independent of a place's physical effects, we can imagine the steps that would get us there, if we wanted.

As such, time constitutes a horizon for possible human actions, something we have also seen in Heidegger. But the universal component of this horizon is most evident in Kant, who identified man's time consciousness with his ability to conceptualize lines 'in thought',[9] despite our locally bound, practical lives. This capacity allows us to formulate ideas about what happens along a timeline drawn from a randomly distant past to an equally randomly distant future. When we humans act in the world, we always relate to past and future as a horizon for such imagined lines, and as I understand Kant, he argues that this horizon constitutes time itself.

If this idea is true, then mankind's particular action consciousness endows McTaggart's A-series with a reality that exceeds dogs' and other animals' situationally determined expectations and memories, without excluding the B-series. As we saw earlier in the chapter, Kant emphasizes that we cannot conceptualize timelines that do not also have spatial extension. Understood as a time series, such lines represent an open future and a closed past of completed facts. Thus we cannot avoid considering the past as a series of events placed side-by-side in space, in other words, as a static B-series. Time is thereby inconceivable without both the A-series and B-series.

In this context, I have no more to say about Kant's special form of time idealism. But I will return to the issue from a different perspective towards the end of the next chapter, which will round out this discussion of time's nature with a subject I have until now only touched upon: both the A- and B-series consist of time units with specific, extended durations, and that at the centre of all this time we find the unit 'now'. I will first look closer at the practical role the now can play in human life. Then I discuss what it signifies that the physical units that comprise the now themselves have a specific duration.

SIX

THE HUMAN *NOW*

'Wait, don't start celebrating . . . *now* he's across the finish line!' When it comes to 'now', everyone knows this can be a matter of hundredths of a second. Or it can be a matter of months and years, as when, after thirty years, you meet an old friend you have not seen since elementary school. 'And you, what are you doing now?' he might ask, and you answer, 'Now I'm married to Anne and I work at Hydro.' Both things have happened in the last twenty years. Nonetheless, it is an adequate answer regarding what you are doing 'now', as opposed to 'before', when you were single and worked at Shell. As we see, the now is an extremely malleable concept, and fantasy alone sets the limits for what a living being in a given situation might perceive of as now. The now's extent, in other words, is situationally determined, or contextual, not fixed by nature once and for all.

The now, meanwhile, contains context-independent, physical units. From situation to situation, it can vary whether by 'now' we mean a second or a year, but that does not imply that the physical content of 'year' or 'second' changes. Traditionally, a year is defined as the Earth's rotation around the Sun, and a second is currently defined as a certain number of caesium atom vibrations.[1] Such physical series form time's physical counting units, and these do not vary from situation to situation. Every physical time unit displays, in contrast, another kind of variety, namely, our subjective experience of how long it lasts. This is also an idea whose limits we can scarcely imagine, something, I will argue at the end of the chapter, that shows that time itself is essentially subjective.

Meanwhile, I will first take a closer look at the now in contrast to the past and the future. In the previous chapter, we became acquainted with what we can call theoretical presentism: only that which is contained in a 'now' is truly real. What the now's contextual variance has to say about this type of presentism I do not fully know, and I will not pursue the subject further here. In any case, there also exists a practical presentism, which concerns being present in all that one is doing at any moment, or, as the catchphrase has it, 'living in the now'. This is something that has concerned many authors, philosophers and other thinkers.

Presentism in Practice

The question of the now's reality and its duration are not only of theoretical interest but concern the basic existential experience regarding the finiteness of things. Nothing in our lives can be saved from time's bite, and is this not a truth that can fill us with the blackest despair? It is frustrating enough that Things Take Time. Is it not worse still that Things Are Lost? Fernando Pessoa lets us partake in his experience of the tragedy of loss:

> Time! The past! ... That which I was and will never again be! That which I had and will never again have! The dead!

The dead who loved me in my childhood. Whenever I
remember them, my whole soul shivers and I feel exiled
from all hearts, alone in the night of myself, weeping like
a beggar before the closed silence of all doors.

Past events posing as closed doors we can never open again – confronted by this experience, is it not a pure escape from reality to say, like Augustine and other presentists, that the past is only a memory? Is it not instead the past *as* past from which we are so tragically barred? In order to approach this idea, we must take a closer look at what presentism signifies in literature and in life.

The sorrow that Things Are Lost also weighed on the young Marcel Proust (1871–1922), the author of the multi-volume novel *In Search of Lost Time*.[2] Love, the good, the beautiful – everything that makes life worth living does not persist but is instead swept into the gloom of the past, and how different a real, present love seems from the remembrance of a lost love! Or is there perhaps a way back? Can old love be reawakened to life? Proust concluded that it could, and while Pessoa wept before the past's locked doors, Proust thought he had found the key to open them. It happened through an artistic perception of the past's tangible presence in the now.

Proust discovered that there are two very different ways of relating to the past. On the one hand, we can recline there and ponder it, recapitulating things that happened during our life or reading about historic events. As with studying a photograph, we will often then experience a significant distance between the present and the past, and a pleasant memory can feel painful. On the other hand, certain sensations can bring to life the experience of the past *as it was*, that is, as being qualitatively the same experience that we had once upon a time. Both in Proust's book and in his life, this re-experience was connected to world literature's most famous snack.

This happened when the fictional character Marcel ate a madeleine as an adult. The taste brought back the pleasant childhood experience of his aunt giving him madeleines after first dipping them in tea. The experience was reawakened not only in a metaphorical sense, and Proust did not just have a memory of once eating the madeleine. He literally re-experienced the taste as it was from back then, not as a memory of a different experience, but as part and parcel of the same experience. After this event, Marcel gave himself over to recapitulating past experiences. It was as if his childhood village of Combray took form and revived to new life, with madeleines, aunt, house, garden, society, church and everything else that came to populate the novel's universe.

The fact that taste and smell can breathe life into a whole past world is itself a general human experience. The Norwegian *Dagbladet* journalist Eirik Alver describes how early one morning, after a night spent in the city, he happened upon a cold cola bottle, only to discover it was not a plastic bottle but a glass bottle like the ones in his childhood. He then put the bottle to his mouth:

The next moment something exceptional happened. As the liquid filled my mouth, I got a feeling so overwhelming that I had to sit on a bench. A new swallow and the same thing occurred. Suddenly, I was back in time. To a time with scrapes and slingshots. New tennis shoes and a new issue of *Tempo*. Aunt Ragne and Uncle Rolf. Pop candy, gum balls, and cola. Cola that tasted like cola should.[3]

But whereas for most of us this is only a sporadic experience, literature for Proust became an enduring, systematic way to revitalize the past. Does this road stand open for all, even beyond art?

Other types of presentism more directly aim at changing human life, as a kind of art of life. The way we live our life today can hinder us experiencing the past as an impenetrable door protecting yesterday's treasures. Then the fantastical, monstrous door guard bursts like a troll at sunrise, so that sorrow over bygone days loses its power over us. The classical Indian poet Kalidasa has shaped a poetic formulation of this viewpoint on past, future and present in his *Hymns of Dawn*. 'Today well lived makes every yesterday a dream of happiness, and every tomorrow a vision of hope,' as the poet says.

If we turn from ancient poetry to philosophy, we find similar thoughts in the Roman Stoic Seneca (*c*. 4 BC–*c*. AD 65). 'On the Shortness of Life' is the title of one of his epistles, referencing the human lament of everything left undone in life. But life is not too short, says Seneca, it is long enough for those who live correctly and do not abandon themselves to restless desire or striving for honour and fame. There is still much that we will never achieve, and everything we do achieve we risk losing. But an attitude of philosophical equanimity is something of which no one can rob us against our will, and for the wise the past is not a collection of lost goods but an eternal source of insight.

The surest way to make life too short is, by contrast, only to plan for the next year, to worry about tomorrow, or to look forward to the day when, finally, you will live. The day in which you should live is *today*, Seneca asserts, and if you want to get as little from life as possible, it will be by spending your best years longing for the day you can leave the employee ranks and escape public duties. Life is long enough for those who live meaningfully in the day, and it is too short for those who always have something they must do first:

> They keep themselves very busily engaged in order that they may be able to live better; they spend life in making ready to live!... The greatest hindrance to living is expectancy, which depends upon the morrow and wastes today. You dispose of that which lies in the hands of Fortune, you let go that which lies in your own. Whither do you look? At what goals do you aim? All things that are still to come lie in uncertainty; live straightway![4]

I have no idea if John Lennon ever read Seneca, but Seneca would give Lennon a knowing smile: 'Life is what happens to you while you're busy making other plans.'

To the extent that this truly is the road to overcoming the tragedy of the past, is it also the road to overcoming time itself? That question is worth posing in light of Augustine's assertion that if something is a part of time, it contains elements of past, future and present. To live in a pure now means living outside of time. And, according to Augustine, only Christian faith can bring us into contact with this timeless, pure present. To God alone belongs eternity, which is identical to a *nunc stans*, a 'standing now', without past and future horizons. But other religions and life views also hold the idea that living fully and wholly in the now is tantamount to making time disappear. A modern advocate for this viewpoint is the German American author of the bestseller *The Power of Now*, Eckhart Tolle (b. 1948).[5]

Tolle is a man who, amid a life crisis, received a sudden, personal revelation. Overwhelmed by anxiety and suicidal thoughts, he awoke one morning to a qualitatively different day. Everything around him was brand new and dew-fresh, as if he were perceiving the world for the first time. With that his life changed, for the experience gave him insight into a personality-altering presentism: only the now is real, it is always now, and if you befriend the now, you feel at home no matter where you are.

In order to befriend the now, Tolle argues that we need a practical rule of life. Are you dissatisfied with your life here and now? Then you have three possibilities. You can do something to escape the situation. You can change the situation. Or, if both are impossible, you must simply accept the situation as it is. The road to happiness is to say yes to life as it is, here and now.[6] Painful situations feel painful, but through acceptance we can at least rid ourselves of the additional burden of painful thoughts: 'If only I were not here!', 'Just think, what if I never get better?' or 'I was an idiot to consign myself to this!'

At the root of Tolle's life attitude, we find a Buddhism-inspired view of the human 'self' or 'ego'. We are not identical with our thoughts and emotions, which always reach out over the now. Deep within, we are one with existence's great void. If we consider the now from that perspective, it is like peering through time and everything that otherwise disrupts our pure access to Being. It is like waking from dreams of past and future and seeing the world as it truly is. The world then appears with a beauty that it is only possible to experience from a position within Nothingness, which is identical to death. If we befriend the now, then death becomes our friend as well.

Is this idea humanly possible? Can you truly make time vanish through adopting a pure time consciousness? I leave this to the reader to judge relative to his or her own life. But that it is to some extent possible we know from the general human experience of time appearing to stand still in the face of strong, intensely meaningful experiences. This demonstrates that the now's duration is limitlessly flexible, not just because it varies contextually with the amount of physical units it contains, but because the experience of these units is so variable.

Subjective Time

Hinduism distinguishes between living, ancestral and divine time. Our physical day from sunrise to sundown corresponds to an ancestral month and a divine year. It is reasonable here to recognize that ancestors and divinities experience a month or a year as we experience a day; the idea should not give us too much trouble. For if there is anything that varies from situation to situation, it is the subjective experience of time's progression.

'Time raced ahead'; 'The night slipped infinitely on'; literature is littered with descriptions of experienced time tempo, descriptions that every reader immediately comprehends. For although each of us has our own personal experiences, a number of common features are also associated with subjective time variation. Just think of how quickly time passes with age. Physically, there is no change in Earth's rotation around the Sun, but how quickly a year elapses at a riper age compared with a year in early childhood! Even a month's summer holiday seemed a small eternity, after all.

We also recognize a common attribute regarding short and long time experiences respectively. If we are bored, we concentrate on time's passage instead of engaging ourselves in something, and so time passes slowly. If we are entertained, we get caught up in the situation and forget time, and so it passes quickly. Things seem to take longer, therefore, when we are passively waiting than when we are actively doing something. I myself have often experienced that while in the shower time seems to fly, though when you're standing outside the bathroom waiting time seems gruesomely slow. And how quickly can the time pass when we are engaged in the best thing we know: being together with the one we love? Equally slow is time's passage for those who wait on their beloved, as Juliet waited on her Romeo:

> I must hear from thee every day in the hour,
> For in a minute there are many days:
> O, by this count I shall be much in years
> Ere I again behold my Romeo![7]

Too bad it is not the other way around, one might say, that the delightful time passes slowly and the tedious quickly. On the other hand, something strange happens when a tedious or delightful present slips into the past, whereupon they seem to exchange roles. Looking back, content-rich, engaged parts of life emerge as long, whereas the tedious, content-poor parts seem short. Surely that is connected to the fact that purely subjective experience gets lost, to be replaced by what is more or less content-rich.

What is then the correct time experience, some might ask? Exactly how long is time, independent of our mood? As long as we are talking about time that can be experienced, the question seems meaningless, something Thomas Mann indicates in *The Magic Mountain*. The feverish Joachim, Hans Castorp's cousin, is about to get his temperature taken, something that invites a few reflections surrounding experienced and measured time:

'Yes, when you pay close attention to it – time, I mean – it goes very slowly. I truly like measuring my temperature four times a day, because it makes you notice what one minute, or even seven, actually means – especially since the seven days of a week hang so dreadfully heavy on your hands here.'

'You said "actually". But "actually" doesn't apply,' Hans Castorp responded. He was sitting with one thigh hiked up on the railing; the whites of his eyes were bloodshot. 'There is nothing "actual" about time. If it seems long to you, then it is long, and if it seems to pass quickly, then it's short. But how long or how short it is in actuality, no one knows.'[8]

No, it is something no one knows, and that scarcely anyone *can* know. For it is one thing to count the minutes. It is another thing to recognize a minute's actual duration, which we experience in such extremely different ways. With accidents and other near-death situations, time's passage can be tremendously slow, so slow it almost seems to stand still. We must also assume that non-human creatures might experience time in a completely different way from us. How long is a day to a cat? To an elephant? To a rattlesnake? To a mayfly? To conscious beings in far galaxies or to supernatural beings and gods? Whether we believe in exotic gods and lifeforms or not, there is no reason to exclude them because they might have 'wrong' time experiences.

And yet, if there are no 'wrong' or 'right' time experiences, then can there be no objective truth regarding the duration of time sequences? In order to approach this idea, I will finally look closer at the relationship between subjective time, physics time and narrative time. It turns out that there is a closer connection between physics time and narrative time than one might think, even though time itself has a unique subjectivity that is captured neither in narrative nor in physics. The reason for this last fact is that time is not a tangible or general phenomenon, but something particular. That is, actual time sequence has a specific duration, not a duration we only imagine.

Time's Human Duration

On a not particularly pleasant day, a man named Alberto was knocked down, taken to an unknown location and imprisoned in a basement belonging to the insane behavioural researcher Alfredo. There he spent the next ten years of his life, supplied by Alfredo with food and drink, who filmed his movements all day. It quickly turned out that a knock to the head during the attack had destroyed Alberto's retention. He still remembered quite a bit of maths and his reading ability was intact. The same was true of important, general truths regarding the shape and appearance of objects. But there are so many strange brain injuries and, in Alberto's case, all concrete knowledge of extended magnitudes in time and space had vanished. Great was his joy, therefore, when he discovered in Alfredo's basement a library containing volumes on physics, cosmology and biochemistry. Without anything else to do but read, after seven or eight years he could have entered a quiz game with the world's leading natural researchers.

One significant knowledge gap, meanwhile, Alberto was not able to bridge before he escaped. The books were written in a disciplinary language without a single reference to things he recognized from his daily activities in the basement. This was particularly true for the duration of time sequences. He read about the caesium atom's fluctuations in relation to the heart rhythms of a number of unfamiliar animals and in relation to the movements of unknown heavenly bodies. But he could not view any of these in relation to his own heartbeat or his own basement wanderings. Therefore, he had no idea how many seconds corresponded to his personal, indexical experience of the things he did now, soon or for a little time back. It was as if theoretical and practical time conceptions were products of different worlds.

How can we best characterize such theoretical knowledge? It originates in books, so from other people's discussions of what is happening in nature around us. From these accounts, Alberto understood that the issue concerned time sequences of a certain duration, although he was still ignorant of the specific duration of things.

As we have seen, something similar can be said about narrative time awareness. In stories, we move freely relative to factual time sequence, which means in thought we overlook the actual, specific time sequence of 'minutes' and 'years'. Disconnected as it is to Alberto's real life, his theoretical time conception is, in a way, pure narrative time. In addition, I believe we can say that it is pure physics time, which is also of undetermined duration. That is, narrative and physics time have something important in common.

'Last year the Earth changed rotation around the Sun and in three years we will see the effects of it,' could be taken from a recent documentary novel. It took me only three or four seconds to narrate three actual years, and in thought this process occurs perhaps even faster. Nonetheless, the distinctive aspect of narrative time does not lie in picturing a world where everything happens extremely fast in relation to the seconds it takes to form a statement. A sentence, after all, enters into the narrated world's course, so the connection between it and the rest of the world we imagine as unchanged. The distinctive aspect is rather that we imagine time series completely independently of their specific duration. What we overlook, that is, is not the relationships of time between things happening in the world, but a global characteristic of that world. The world's unfolding consists of a certain number of time units, and if we overlook the unit's own duration, we overlook the entire world's actual duration. That is something one also does in physics.

When a physicist counts and measures things, he or she assumes numerical units, the quantity of which it is pointless to interrogate. In order to count dogs, we must first know what a dog is, and to count ski jumps we must first know what a ski jump is. If next we want quantitative knowledge regarding a specific number of dogs or of ski jumps, we must either discover how many units they consist of ('How many jumps are in the run?') or compare them with other amounts ('Who has the most dogs, you or your neighbour?') But it is senseless to question how many dogs is 'one dog' or how many ski jumps is 'one jump', as if it were a particular quality of dogs or ski jumps. The same is true of seconds, minutes and other physical time units.

What then is the truth regarding the specific duration of a minute? We will never discover its magnitude among the qualities we sense 'in' a minute. For magnitudes are not qualities, they are quantities, and quantities can be determined only through counting or comparing them to each other as we do when we compare other events to the minute hand's movement on the dial.

Alberto did exactly that. We imagine certainly that the books he read did not deal with fictive scenarios, but actual, observed, physical series. And he knew that these series had the extended property of duration. But he was unable to make any more specific determination regarding the length of the duration. Even though he knew more than most of us about described time sequences, he lacked specific knowledge about what most of us take for granted. Meanwhile, this is not considered a lack within physics, where the length of a time sequence is consequently treated as the relationship between time sequences, and the actual magnitude of time units is overlooked. As we have seen, there are good reasons to do this; indeed, it seems almost pointless to pursue facts regarding the 'inner' magnitude of time units.

Nonetheless, when it comes to time's subjectivity, the question is meaningful to us. Experienced duration varies with our activities, and these variations do not concern the relationship between physical events but the magnitude of the individual time unit. Just think how slowly time can pass at the station before the train arrives to take you to your job on a cold winter morning, or how quickly it can pass before you must say farewell to your loved one at that same station on a warm summer evening. It is not that the station clock or the train's movements are experienced as slow or fast in relation to other things. *Everything* is experienced as slow, or fast. That is, subjective time is global; we always experience a definite and not just an imagined duration from *now* until *now*. But a corresponding, objective property of time units simply does not exist.

The conclusion here becomes that, without the potential time experience of various human actions, physical events have only a perceived, general duration. If we further maintain that time sequences have a particular, not only a conceptual, general duration, then time becomes inextricably connected to the existence of human, acting subjects. It is tempting to say that time is us! Or, a little more precisely, as in the previous chapters, that time is the human past-and-future horizon that surrounds our potential present actions.

This perspective provides a simple explanation for a fact that many time philosophers have pondered: that we cannot just move around in time as we do in space. Time merely ticks and flows away and carries us with it whether we will or no. This impotence can be regarded as an expression of a missing ability to control an objective process. But picture time's river as a violent flood that no person can escape once they have tumbled in. In that case, our ability to move around in space means that it is, in a way, even more subjective than time. At least it does not have the power over us that time has.

However, I want to believe that the opposite viewpoint is just as plausible: if we cannot move around in time, it is because time is the thorn in our side we can never escape because we *are* time. Time is not a river, it is us, as the Argentine author Jorges Luis Borges (1899–1986) puts it:

> Time is the substance of which I am made. Time is a river that sweeps me along, but I am the river; it is a tiger that mangles me, but I am the tiger; it is a fire that consumes me, but I am the fire. The world, unfortunately, is real; I, unfortunately, am Borges.[9]

If that is how reality is, both for Borges and for us, then we have caught sight of a philosophical answer to the philosophical question concerning time's essence. In the best Socratic tradition, we can indicate mankind's place in the world – and perhaps even say that man's reflection over time is time's self-reflection.

But if time is something human, what of the dinosaurs and all the other creatures who existed before we appeared on the earth? Did things exist or did they not exist prior to us? This question I will only answer with, 'Yes indeed, to both,' with reference to two distinct roles for the human body. If we are time, we embody it as active subjects, and as living organisms we are also spatial objects. The last idea corresponds to the fact that time, as we have seen, is inconceivable without spatialization. If we, therefore, as subjects, have a horizon of possible actions, we are also located as objects before and after other things set within that same horizon. If we are the river, still it draws us on.

REFERENCES

INTRODUCTION: WHICH TIME?

1 Roland Barthes, *Camera Lucida: Reflections on Photography*, trans. Richard Howard (New York, 1982), pp. 18, 89.
2 Augustine, *The Confessions*, trans. Maria Boulding, OSB, ed. John E. Rotelle (New York, 1997), p. 295.
3 Fernando Pessoa, *The Book of Disquiet: The Complete Edition*, trans. Margaret Jull Costa (New York, 2017), p. 384.
4 With modifications from Pope Gregory XIII's calendar reform in 1582.

1 THE CLOCK AND ITS PAST:
ON TRADITIONAL AND MODERN TIME CONCEPTIONS

1 Anders Johansen, *All verdens tid* (Oslo, 2001), p. 10.
2 Robert Levine, *A Geography of Time* (New York, 1997), Chapter Four. Regarding the contrast to modern, social 'time use', see Robert H. Lauer, *Temporal Man: The Meaning and Uses of Social Time* (New York, 1981), pp. 86ff.
3 Johansen, *All verdens tid*, p. 88.
4 Ibid., pp. 85f.
5 More on this subject in Chapter Six.
6 Johansen, *All verdens tid*, p. 89.
7 Jonathan Swift, *Gulliver's Travels* (Ware, 1992), p. 24.
8 Julio Cortázar, *Cronopios and Famos*, trans. Paul Blackburn (New York, 1999), pp. 23–4.
9 The Rolling Stones, 'Time Waits for No One' (1974).
10 Max Weber, *The Protestant Ethic and the Spirit of Capitalism*,

trans. Talcott Parsons (London and New York, 1992), p. 104. Footnotes excluded.

11 Thomas Hylland Eriksen, *Small Places, Large Issues: An Introduction to Social and Cultural Anthropology*, 2nd edn (London, 2001), pp. 251–2.

12 Johansen, *All verdens tid*, pp. 97f.

2 PHYSICS TIME: EINSTEIN

1 Galileo Galilei, *Dialogue Concerning the Two Chief World Systems*, trans. Stillman Drake, ed. Stephen Jay Gould (New York, 2001), pp. 216–17.

2 A strikingly simple representation of this idea can be found here: http://casa.colorado.edu/~ajsh/sr/paradox.html. (Don't forget to go to the 'Center of the Lightcone' further down the web page.)

3 I will return to this idea towards the end of Chapter Six.

4 Rudolf Carnap, 'Intellectual Autobiography', in *The Philosophy of Rudolf Carnap*, ed. P. A. Shilpp (La Salle, IL, and London, 1963), pp. 37f.

5 From Einstein's letter to Besso's son and sister, 21 March 1955, cited in Helen Dukas and Banesh Hoffmann, *Albert Einstein, Creator and Rebel* (New York, 1972), pp. 257f. For a different view on relativity and death, see my 'Time, Death, and Duration', forthcoming in *Zeitschrift für philosophische Forschung* (2019).

3 PHILOSOPHY'S TIME: BERGSON, HUSSERL, HEIDEGGER

1 Robert Geroch, *General Relativity from A to B* (Chicago, IL, and London, 1978), pp. 20f.

2 Paul Davies, 'That Mysterious Flow', *Scientific American*, CCLXXXVII/3 (September 2002).

3 Henri Bergson, *Time and the Free Will: An Essay on the Immediate Data of Consciousness*, trans. F. L. Pogson (Mineola, NY, 2001), p. 100.

4 Cf. p. 24.

5 Here I am reading Bergson's philosophy in a way that is intended to make the most sense, and one might well call it a benevolent interpretation. Those who wish for a more hostile approach will thoroughly enjoy Bertrand Russell, 'The Philosophy of Bergson', *The Monist*, XXII (1912), pp. 321–47.

6 This question is open to debate. In terms of background literature, I recommend Dominic Pedler, *The Songwriting Secrets of The Beatles* (London and New York, 2003), Chapter Thirteen (pp. 475–516):

'The "A Hard Day's Night" Chord: Rock's Holy Grail'. The chapter provides a thorough overview of the research around this chord.

7 Not just as phases of each sense, but as what he sometimes calls 'distant' retentions, that is, in the relation between senses.

8 Maurice Merleau-Ponty, *Phenomenology of Perception*, trans. Donald A. Landes (Abingdon and New York, 2012), p. 446 (485).

9 In any case, if we believe Carol J. White, *Time and Death: Heidegger's Analysis of Finitude* (London, 2005).

10 A thorough argument for this last viewpoint is found in William D. Blattner, *Heidegger's Temporal Idealism* (Cambridge, 1999).

4 RECURRENCE TIME IN LIFE, RELIGION, HISTORY AND LITERATURE

1 Among other places, this idea is presented in Sigmund Freud, *Beyond the Pleasure Principle*, trans. Gregory C. Richter (Toronto, 2011).

2 This thought is put forward in Friedrich Nietzsche, *Thus Spake Zarathustra: A Book for Everybody and Nobody*, trans. Graham Parkes (Oxford, 2005).

3 See Kierkegaard's literary work *Repetition* in *Repetition and Philosophical Crumbs*, trans. M. G. Piety (Oxford, 2009).

4 See Mircea Eliade, *The Sacred and The Profane*, trans. Willard R. Trask (Orlando, FL, 1957), pp. 78f.

5 Reinhart Koselleck, *Futures Past: On the Semantics of Historical Time*, trans. Keith Tribe (New York, 2004). For Augustine, this idea is tied to the fact that time, as we shall see in Chapter Five, is located only in consciousness, independent of external, worldly realities.

6 Ibid., pp. 9f.

7 Ibid., pp. 33f.

8 See pp. 79f.

9 Alasdair MacIntyre, *After Virtue: A Study in Moral Theory* (London, 1981), p. 215.

10 Ibid., p. 216.

11 Paul Ricoeur, *Time and Narrative*, trans. Kathleen McLaughlin and David Pellauer (Chicago, IL, and London, 1990).

12 Ibid., p. 67.

13 Ibid.

5 IS THERE ANY TIME OUT THERE?

1 For Hegel, however, this idea did not lead to a denial of time's reality, 'Time is the Concept itself that *is there*', in *The Phenomenology of Spirit*, trans. A. V. Miller (Oxford, 1977), p. 801.
2 Thomas Mann, *The Magic Mountain*, trans. John E. Woods (New York, 2005).
3 Cf. Kant, *Critique of Pure Reason*, trans. Paul Guyer and Allen W. Wood (Cambridge and New York, 1998), B50.
4 G. W. Leibniz, *New Essays on Human Understanding*, trans. Peter Remnant and Jonathan Bennett (Cambridge, 1996), p. 138.
5 Cf. the beginning of Chapter Two.
6 Here we recognize some of Bergson's points, cf. pp. 65f.
7 Cf. William D. Blattner, *Heidegger's Temporal Idealism* (Cambridge, 1999), pp. 261–71.
8 See p. 58.
9 See, for example, Kant, *Critique of Pure Reason*, B154–6.

6 THE HUMAN *NOW*

1 More precisely, as the time it takes for a caesium-133 atom to vibrate 9,192,631,770 times.
2 Also translated in English as *Remembrance of Things Past*.
3 Eirik Alver, 'Klinkende glassklart' (Clear as Glass), in *Dagbladet*'s 'Magasinet', 8 May 2010.
4 Lucius Annaeus Seneca, 'On the Shortness of Life', trans. John W. Basore (Los Angeles, CA, 2017), p. 22.
5 Eckhart Tolle, *The Power of Now: A Guide to Spiritual Enlightenment* (Novato, CA, 1999).
6 Cf. the discussion of Nietzsche, pp. 88f.
7 William Shakespeare, *Romeo and Juliet*, http://shakespeare.mit.edu/romeo_juliet/full.html.
8 Thomas Mann, *The Magic Mountain*, trans. John E. Woods (New York, 2005), pp. 75–6.
9 Jorge Luis Borges, 'A New Refutation of Time', trans. Suzanne Jill Levine, *The Total Library: Non-fiction, 1922–1986*, ed. Eliot Weinberger (London, 1999), p. 332.

FURTHER READING

INTRODUCTION: WHICH TIME?

Barthes, Roland, *Camera Lucida: Reflections on Photography*, trans. Richard Howard (New York, 1982). The book is a fascinating, short study of photography's time and timelessness.

1 THE CLOCK AND ITS PAST

Johansen, Anders, *All verdens tid* (Oslo, 2001). A treasure of a book on the subject of man's relationship to time in modern and premodern societies. Unfortunately, it is not yet available in English.

Landes, David S., *Revolution in Time: Clocks and the Making of the Modern World* (Cambridge, 1983). A detailed description of the mechanical clock's history and social significance.

Levine, Robert, *A Geography of Time* (New York, 1997). Also a book highly worth reading on different time conceptions.

Marx, Karl, *Capital*, vol. I, trans. Ben Fowkes, ed. Friedrich Engels (London, 1990)

Weber, Max, 'Protestant Asceticism and the Spirit of Capitalism', in *Selections in Translation*, trans. Eric Matthews, ed. W. G. Runciman (Cambridge, 1978)

2 PHYSICS TIME: EINSTEIN

Feynman, Richard P., *Six Not-so-easy Pieces: Einstein's Relativity, Symmetry, and Space-Time* (London, 1997). This book contains portions of the renowned *Feynman Lectures on Physics*.

Galison, Peter, *Einstein's Clocks, Poincaré's Maps: Empires of Time*

(New York, 2003). This book describes the connection between Einstein's work at the Bern patent office and the development of the theory of relativity.

Kennedy, J. B., *Space, Time and Einstein: An Introduction* (Chesham, 2003). A clear and simple introduction to the natural scientific aspects of time-and-space philosophy, viewed in relation to Einstein.

Wyller, Truls, *The Size of Things* (Paderborn, 2010). Chapter Two provides a more extensive discussion of the special relativity theory than the one I have provided above.

3 PHILOSOPHY'S TIME: BERGSON, HUSSERL, HEIDEGGER

Bergson, Henri, *Time and the Free Will: An Essay on the Immediate Data of Consciousness*, trans. F. L. Pogson (Mineola, FL, 2001)

Heidegger, Martin, *Being and Time*, trans. Joan Stambaugh (Albany, 2010). Even if one does not read a work from cover to cover, small samplings also have their value.

Husserl, Edmund, *On the Phenomenology of the Consciousness of Internal Time (1893–1917)*, trans. John B. Brough (Dordrecht, 1991). Volume X of the author's collected works, *Husserliana*, is available here.

Polt, Richard, *Heidegger: An Introduction* (New York, 1999). Perhaps the best introduction to his work.

Russell, Matheson, *Husserl: A Guide for the Perplexed* (London and New York, 2006). An excellent introduction to Husserl's thought.

4 RECURRENCE TIME IN LIFE, RELIGION, HISTORY AND LITERATURE

Eliade, Mircea, *The Sacred and The Profane*, trans. Willard R. Trask (Orlando, FL, 1957)

Gadamer, Hans-Georg, *Truth and Method*, revd trans. Joel Weinsheimer and Donald G. Marshall (London, 2013)

Koselleck, Reinhart, *Futures Past: On the Semantics of Historical Time*, trans. Keith Tribe (New York, 2004)

Nietzsche, Friedrich, *Thus Spake Zarathustra: A Book for Everybody and Nobody*, trans. Graham Parkes (Oxford, 2005)

Ricoeur, Paul, *Time and Narrative, Volume I*, trans. Kathleen McLaughlin and David Pellauer (Chicago, IL, and London, 1984). The most important section is the first part of Volume I.

Warnke, Georgia, *Gadamer: Hermeneutics, Tradition, and Reason* (Stanford, CA, 1987). A good, explanatory overview of Gadamer's authorship.

5 IS THERE ANY TIME OUT THERE?

Augustine, *The Confessions*, trans. Maria Boulding, OSB, ed. John E. Rotelle (New York, 1997), Book 11. One of European time philosophy's most important texts.

Le Poidevin, Robin, *Travels in Four Dimensions: The Enigmas of Space and Time* (Oxford, 2003), Chapter Eight. A more comprehensive presentation of McTaggart.

McTaggart, John, 'The Unreality of Time', *Mind*, XVII (1908), pp. 457–74. The article is an uncommonly dense piece of philosophy that I would not recommend someone tackle outright. But it is also found in a number of anthologies and can be good to have on hand if you are reading one of the many introductions on the subject, for example, J. B. Kennedy's book *Space, Time and Einstein*, mentioned in Chapter Two.

Mann, Thomas, *The Magic Mountain*, trans. John E. Woods (New York, 2005). If you were to choose two 'novels of time' from the previous century, this would be one of them. The other is discussed in the next chapter.

Searle, John, *Rationality in Action* (Cambridge, MA, 2001). An interesting discussion of the time dimensions surrounding human statements, promises and contracts.

6 THE HUMAN *NOW*

Proust, Marcel, *In Search of Lost Time*, trans. C. K. Scott Moncrieff and Terence Kilmartin (New York, 2003). The previous century's second great novel of time.

Seneca, Lucius Annaeus, 'On the Shortness of Life', trans. John W. Basore (Los Angeles, CA, 2017).

Wyller, Truls, 'How Big? How Fast? Transcendental Reflections on Space, Time and World Models', *Philosophy*, LXXXIV/3 (2009), regarding the difference between the conceptual and the actual, particular dimension of things in time and space. See also 'What Peter Didn't Know', in *Zeitschrift für philosophische Forschung*, LXXII/4 (2018), pp. 502–510.

INDEX

A-series (McTaggart) 114, 116–17, 120, 124–5

abstract time 23, 38, 40

action time 19–34, 39–40

after *see* before, after, simultaneously

Altdorfer, Albrecht 93

Alver, Eirik 129

Aristotle 102–3

art 16, 37, 97–8, 101, 128

Armstrong, Neil 43, 45–6, 56

Augustine, St 9, 92, 107, 112–13, 117, 122, 128, 131

B-series 114–16, 124–5

Barthes, Roland 8

Beckett, Samuel 81

before, after, simultaneously 58, 60, 114

Bergson, Henri 16, 62–71, 73, 81, 85, 118, 120

Besso, Michele 61

Blattner, William D. 144, 145

Borges, Jorge Luis 140

Buddhism 132

Caesar, Julius 12

calendar 11–12, 65–6

Carnap, Rudolf 59–60, 143

change 7, 10–13, 17, 41, 48, 50–51, 62–4, 70–71, 78, 80, 87, 89, 97, 109, 114–22

Christ 91

Christianity 37, 91–2

Chronos 10

clock 11–12, 14–15, 18–34, 39–42, 43–4, 47–8, 65–6, 71, 112, 115, 139

concrete time 29–30, 38–9, 107, 136

configured 102

continual time 24

Copernicus, Nicolaus 49

Cortázar, Julio 32

Davies, Paul 143

death 8, 13, 37, 61, 84, 88, 99, 102, 132, 135

discrete time 22, 24

duration 17, 20–22, 28, 57, 59, 65, 108–9, 112–13, 125, 127, 132, 135–40

durée (Bergson) 65

ecstatic (Heidegger) 78, 80, 82–3, 85, 99
Einstein, Albert 15–16, 43–61, 63, 114
Eliade, Mircea 90
eternity 37, 88, 90, 92, 131
event time 19, 21, 24, 26
existential time 10, 80, 82, 88, 99, 127
expectations 74, 113, 122–4
extension 108–13, 124

Feynman, Richard 146
finitude 85
Franklin, Benjamin 36–7, 40
Freud, Sigmund 87–8
future see past

Gadamer, Hans-Georg 16, 97–8, 102
Galilei, Galileo 15, 49
Galison, Peter 146
general time 94, 103, 136, 139
Geroch, Robert 143

Hegel, Georg Friedrich Wilhelm 106, 145
Heidegger, Martin 16, 64, 71, 78–85, 93, 97, 99, 118, 120, 124
hermeneutics 16, 94, 96, 99, 102
history 13, 16, 30, 41, 76, 85, 86, 92–101, 118
homogenous time 30, 48
horizon (time-) 9, 14, 16–17, 58, 80, 83–4, 93, 96, 118, 120, 124, 131, 140–41
Husserl, Edmund 16, 61, 64, 71–9, 81, 85, 118, 120
Hylland Eriksen, Thomas 39–40

impression (Husserl) 74, 77, 120–21

indexical 121–2, 137
intentionality 72

Janus 10
Johansen, Anders 26, 40
Juliet 134

Kalidasa 130
Kant, Immanuel 107–8, 124
Kennedy, J. B. 147
Kierkegaard, Søren 89
Koselleck, Reinhart 92–4, 97

Landes, David S. 146
Lauer, Robert H. 142
Le Poidevin, Robin 148
Leibniz, Gottfried Wilhelm von 112, 118
Lennon, John 131
Levine, Robert 142
lifeworld 71–2, 76, 81
lightcone 55–6
Lillo-Stenberg, Lars 86
linear time 92–3

MacIntyre, Alasdair 100–102
McTaggart, John 17, 106–7, 113–18, 124
Märtha Louise 52, 54, 56
Mann, Thomas 107, 110, 134
Marx, Karl 35, 37–8
Merleau-Ponty, Maurice 71, 78–80
mimesis 102
mood 67, 81–5
movement 9, 26–9, 49–50, 52–5, 62–4, 75, 85, 92, 95–6, 99–100, 110, 113, 137–9
Mozart, Wolfgang Amadeus 97–8

narrative time 16, 99–105, 109, 136–7
Newton, Isaac 15, 44–6, 56
Nietzsche, Friedrich 88–9, 91
now, the 13, 17, 30, 60, 64, 83, 85, 90, 99, 108, 113, 116, 125–8, 131–3

Parmenides 106
particular time 122, 136, 139
past, future, present 10, 13–14, 16–17, 20, 28, 30, 39, 54, 58–61, 64–85, 89–101, 104, 108–9, 111–19, 122–4, 127–34, 140
Pedler, Dominic 143
Pessoa, Fernando 9, 127–8
physics time 14–16, 41–61, 63–4, 73, 95, 114, 120, 135–8
Plato 11
plot 103–4
Plotinus 107, 118, 122
Polt, Richard 147
possibility 80, 117
prefigured 102
present *see* past
pre-understanding 98
presentism 113, 117, 127–32
Princip, Gavrilo 94
profane time 90–91
progress 93, 96
promise 119, 123
protention 74, 77
Proust, Marcel 128–30

qualitative time 12, 21–4, 29, 66, 71
quantitative time 20–24, 26, 31, 40, 57, 65–6, 138

reality 80, 104, 106, 110, 112–14, 116–18, 122, 124, 127, 140

recurrence 16, 86–91, 94, 104
reference body 51–2, 57
refigured 102
relative time 46, 49–58, 80, 111, 114–16, 120
retention 74–7, 120–21, 136
Ricoeur, Paul 16, 102–5
Romeo 134
Russell, Bertrand 143
Russell, Matheson 147

sacred time 90–91
Sartre, Jean-Paul 71, 83
Saturn 10
Searle, John 148
seconds 12, 20, 126, 137–8
secular time 90, 92–4, 96
Seneca, Lucius Annaeus 130–31
sequential time 63, 79–80, 101
simultaneity 15, 22–3, 42–6, 56, 58
Socrates 11
Socratic 140
spatialization of time 66, 141
subjective time 28, 60, 63, 122, 127, 133–5, 139–40
succession 73, 75–7, 90–94, 114
Swift, Jonathan 30
synchronization 33–4, 42–4, 47–8

time idealism 107, 118, 124
time measurements 47, 110–12
time nihilism 106–7, 112, 118
time units 30, 35, 109, 112–13, 125, 127, 138–9
time's river 78, 140–41
Tolle, Eckhart 131–2

universal time 17, 23–4, 28–9, 40–41, 43–8, 58, 124

Vishnu 10

Warnke, Georgia 148
Weber, Max 35–7

White, Carol J. 144
Wittgenstein, Ludwig 71
Wyller, Truls 147, 148